Big Data Analytics in Oncology with R

Big Data Analytics in Oncology with R serves the analytical approaches for big data analysis. There is huge progressed in advanced computation with R. But there are several technical challenges faced to work with big data. These challenges are with computational aspect and work with fastest way to get computational results. Clinical decision through genomic information and survival outcomes are now unavoidable in cutting-edge oncology research. This book is intended to provide a comprehensive text to work with some recent development in the area.

Features

- Covers gene expression data analysis using R and survival analysis using R
- Includes bayesian in survival-gene expression analysis
- Discusses competing-gene expression analysis using R
- Covers Bayesian on survival with omics data

This book is aimed primarily at graduates and researchers studying survival analysis or statistical methods in genetics.

Big Data Analytics in Oncology with R

Atanu Bhattacharjee

CRC Press
Taylor & Francis Group
Boca Raton London New York

CRC Press is an imprint of the
Taylor & Francis Group, an **informa** business
A CHAPMAN & HALL BOOK

First edition published 2023
by CRC Press
6000 Broken Sound Parkway NW, Suite 300, Boca Raton, FL 33487-2742

and by CRC Press
4 Park Square, Milton Park, Abingdon, Oxon, OX14 4RN

CRC Press is an imprint of Taylor & Francis Group, LLC

Library of Congress Cataloging-in-Publication Data

Names: Bhattacharjee, Atanu, author.
Title: Big data analytics in oncology with R / Atanu Bhattacharjee.
Description: First edition. | Boca Raton : Chapman & Hall/CRC Press, 2023. | Includes bibliographical references and index. | Summary: "Big Data Analytics in Oncology with R serves the analytical approaches for big data analysis. There is huge progressed in advanced computation with R. But there are several technical challenges faced to work with big data. These challenges are with computational aspect and work with fastest way to get computational results. Clinical decision through genomic information and survival outcomes are now unavoidable in cutting-edge oncology research. This book is intended to provide a comprehensive text to work with some recent development in the area"-- Provided by publisher.
Identifiers: LCCN 2022034791 (print) | LCCN 2022034792 (ebook) | ISBN 9781032028767 (hardback) | ISBN 9781032028774 (paperback) | ISBN 9781003185598 (ebook)
Subjects: MESH: Medical Oncology--statistics & numerical data | Algorithms | Big Data | Data Interpretation, Statistical | Programming Languages
Classification: LCC RC263 (print) | LCC RC263 (ebook) | NLM QZ 26.5 | DDC 616.99/4--dc23/eng/20221115
LC record available at https://lccn.loc.gov/2022034791
LC ebook record available at https://lccn.loc.gov/2022034792

ISBN: 978-1-032-02876-7 (hbk)
ISBN: 978-1-032-02877-4 (pbk)
ISBN: 978-1-003-18559-8 (ebk)

DOI: 10.1201/9781003185598

Typeset in CMR10
by KnowledgeWorks Global Ltd.

Publisher's note: This book has been prepared from camera-ready copy provided by the authors.

I dedicated this book to my family Chaitry, Atrideep, Ma and Baba.

Contents

Preface

Over the past decades, computational flexibility has extended the application of the statistical methodology as a handy method in research practice. The oncology disease complexity and existing challenge to obtain the best therapeutic effect has provided immense opportunity to develop the statistical methodology. Perhaps, the acceptance of the gene expression survival analysis in oncology research is relatively low. The adoption rate of the extended statistical methodology through survival gene expression is not much in oncology research practice. This book aimed to make a single source of information on statistical methodology for oncology research to cover several dimensions of omics data with survival analysis. The idea is to overall extending the survival gene expression methodology in oncology research practice. This book is structured with nine chapters to extend Big Data Analytics in Oncology with R. In oncology, Survival analysis is inevitable. The Introduction are proposed to bring in Chapter I. The importance of statistics in the oncology domain is first felt by the analysis of survival analysis. This chapter is focused on conventional survival analysis. It explained that the Cox proportional hazard assumption, testing of hazard assumption, and their relevance in oncology therapeutic area and application using R. In the current context the application of placebo-controlled trial is overruled by inferiority trial due to ethical constraints and availability of several molecules in the oncology domain. The third chapter will be illustrating parametric survival analysis with gene expression data. It is not always difficult to perform parametric survival with Big size omics data. Oncology is a complex disease, and sometimes the application of parametric survival helps to understand the disease progression and thereafter control it. The Statistical methodology for different survival models with parametric approaches are demonstrated in Chapter 3. The fourth chapter develops and presents the competing risk, i.e., cancer patient died due to other causes than cancer is a new issue in oncology data. The decision about the biomarker in the presence of competing for risk is always challenging. This chapter will make a presentation on competing risk analysis to explore relevant biomarker from omics data. The fifth chapter is detailed on biomarker thresholding techniques. Different statistical methodology, thresholding determination and illustration with R.

The sixth chapter is about high dimensional survival data analysis. The conventional technique on survival analysis is a frequentist-based approach and method with Kaplan-Meier and Cox proportional hazard models always performed. However, there is minimal literature support about the Bayesian counterpart on Cox proportional hazard modeling. This chapter will showcase about Bayesian survival analysis using R. Chapter seven will show the definition and approaches about high dimensional data. The R software will be utilized with real oncology data. Different challenges, variables selection methods, briding the survival data with high dimensional data, covariance structure analysis will be elaborated in this chapter using R.

The concept of frailty provides a convenient way of introducing unobserved heterogeneity and associations into models for survival data. In its purest form, frailty is an unobserved random proportionality factor that modifies the hazard function of an individual or related individual. Chapter eight will be prepared on frailty data analysis, theories about it, the application using R. The statistical methodology with time-course gene expression data will be extended in chapter nine. Time-course data analysis in oncology research is suitable due to their ability to consider different time point information for statistical growth curve modeling. The step-by-step illustration of different time point analysis are presented. The results will be compared between the conventional maximum likelihood estimation (MLE) methods. The Bayesian counterpart has been expanded with time-course data analysis using R. A comprehensive statistical methodology is illustrated in chapter eight. The author employed different statistical models through R software to analyze the time course gene expression data. The joint time-course and survival outcome is essential. It helps to accommodate the dependence between two types of responses obtained by two different approaches. Chapter nine is dedicated with examples of the oncology high dimensional data on time-course protein expression of having survival outcomes.

Author

Atanu Bhattacharjee is now a Lecturer in Medical Statistics at Leicester Real-world Evidence Unit, Leicester University, United Kingdom. He worked earlier as an Assistant Professor at the Section of Biostatistics, Centre for Cancer Epidemiology, Tata Memorial Centre, India. He previously taught Biostatistics at the Malabar Cancer Centre Kerala, India. He completed his Ph.D. at Gauhati University, Assam, India on Bayesian Statistical Inference. He is an elected member of the International Biometric Society (Indian Region). He has published over 250 research articles in various peer-reviewed journals.

Symbol Description

$\Delta(t)$	Hazard rate in survival analysis.
$h(t)$	hazard function.
$\Delta_c(t)$	The cumulative hazard.
C	Censoring time.
T	Survival time.
α	is fixed with probability of rejecting the null hypothesis.
β	Type II error.

Chapter 1

Survival Analysis

1.1 Introduction

Survival analysis works with failure time data. It comes with the Kaplan-Meier estimator, log-rank test, and Cox proportional hazard model. However, survival analysis is different from conventional statistical analysis. In traditional statistical analysis, parameters are assumed with cumulative distribution function (CDF). In contrast, the hazard is used in survival data analysis. Survival analysis works with a censoring mechanism, but it is not usual in conventional statistical analysis. Like non-parameter, estimators are easy to adopt in survival analysis by Kaplan-Meier estimators.

Definition 1 (Survival Analysis) *is a collection of statistical procedures for data analysis for which the outcome variable of interest is time until an event occurs.*

Definition 2 (By the time) *It refers to the age of an individual when an event occurs by years, months, weeks, and days.*

Definition 3 (By event) *we mean death, disease incidence, relapse from remission, recovery (e.g., return to work), or any designated experience of interest that may happen to an individual. In survival analysis, we usually refer to the time variable as survival time because it gives them time that an individual has "survived" over some follow-up period.*

T *denotes the response variable (time),* $T \geq 0$.

S(t) The survival function and

$$S(t) = Pr(T > t) = 1 - F(t) \tag{1.1}$$

DOI: 10.1201/9781003185598-1

How survival data?

library(KMsurv)

data(alloauto)

time =Time to death or relapse(months)

event=Leukemia-free survival indicator=δ (0=alive without relapse, 1=dead or relapse)

type=Type of transplant(1=allogeneic, 2=autologous)

library(KMsurv)

data(alloauto)

alloauto[1:5,]

Listing 1.1: How alloauto data looks like?

```
library(KMsurv)
data(alloauto)
head(alloauto)
alloauto
         time    type  delta
1        0.030    1      1
2        0.493    1      1
3        0.855    1      1
4        1.184    1      1
5        1.283    1      1
6        1.480    1      1
7        1.776    1      1
8        2.138    1      1
9        2.500    1      1
10       2.763    1      1
```

Example with allograft data

library(KMsurv)
data(allograft)
patient= Patient ID
time =Time to graft rejection(days)
rejection=Indicator of graft rejection (1=yes, 0=no)
match=Good HLA skin match (1=yes, 0=no)
library(KMsurv)
data(allograft)
allograft[1:5,]

Listing 1.2: How allograft data looks like?

```
  patient time rejection match
1       1   29         1     0
2       1   37         1     1
3       2    3         1     0
4       2   19         1     1
5       3   15         1     0
6       3   57         0     1
```

1.2 Hazard Function

Suppose that $F(t)$ is the cumulative density function of the random variable X of interest. The cumulative hazard function is presented by

$$\Delta(t) = \int_{-\infty}^{t} \frac{dF(s)}{1 - F(s-)} \tag{1.2}$$

This definition is equally effective for continuous and discrete variable with cumulative density function (CDF) of $F(t)$; for X, those can take negative values. If $F(t)$ becomes binary, then Stieljes integration can be performed. If CDF is continuous, then the integration of the right-hand part can be obtained as $-log(1 - F(t))$. So,

$$\Delta(t) = \int_{-\infty}^{t} \frac{f(s)}{1 - F(s)} ds \tag{1.3}$$

and

$$\frac{\delta}{\delta t}\Delta(t) = \frac{f(t)}{1 - F(t)} \tag{1.4}$$

The hazard function $h(t)$ is presented as

$$h(t) = \frac{f(t)}{1 - F(t)} \tag{1.5}$$

Relation between $\Delta(t)$ and $h(t)$ becomes the CDF of $F(t)$ as, $\Delta(t) = \int_{-\infty}^{t} h(s)ds$ and $\frac{\delta}{\delta t}\Delta(t) = h(t)$. Now the conditional probability of the random variable takes value between $[t, t + dt)$ as

$$h(t)dt = \frac{f(t)dt}{1 - F(t-)} = P(t \leq X < t + dt | X \geq t) \tag{1.6}$$

Present the density function $f(t)$ as

$$f(t)dt = P(t \leq X < t + dt) \tag{1.7}$$

This hazard function $h(t)$ could be non-negative with upper bound. Now the cumulative hazard function Δ becomes non-negative and non-decreasing. It can be observed as $\Delta(t) = -\log(1 - F(t))$. Or

$$\Delta\Delta(t_{last}) = \Delta\Delta(t^*) = \frac{\Delta f(t^*)}{1 - F(t^* - 1)} = 1 \tag{1.8}$$

Or,

$$\Delta F(t^*) = F(t^*+) - F(t^*-) = 1 - F(t^*-) \tag{1.9}$$

The inverse formula presented with the CDF of the cumulative hazard as

$$1 - F(t) = e^{-\Delta(t)} \tag{1.10}$$

$$1 - F(t) = \prod_{s \leq t} (1 - \Delta\Delta(s)) \tag{1.11}$$

Now $\Delta\Delta(s) = \Delta(s+) - \Delta(s-)$.

$$1 - F(t) = e^{-\Delta_c(t)} \prod_{s \leq t} (1 - \Delta\Delta(s)) \tag{1.12}$$

where $\Delta_c(t)$ is the continous part of the cumulative hazard:

$$\Delta_c(t) = \Delta(t) - \sum_{s \leq t} \Delta\Delta(s) \tag{1.13}$$

1.3 Censoring

Survival duration is said to be right-censored at time t if it is only observed as more significant than t. Survival time cannot simply be analyzed by continuous outcomes. It involves outcome events. So conventional logistics regression is not suitable enough to work with survival data. If we know some information about the patients but do not know the exact event time, then the censoring is raised. Censoring must be independent of the survival mechanism. Commonly, censoring occurs if a person lost-to-follow-up during the study or person withdraws from the study. The study duration is fixed, and a person does not observe the event while the study ends. The study concludes while the presumed number of events occurred toward the completion of the study. Suppose the random variable T presents the time to death due to different causes. Let C show the time to censoring. It is observed that (X, δ), where $X = T \wedge c$ and $\delta = I[T \leq C]$ with $I[.]$ Now T has density f and C has a continuous distribution function G.
If the exact survival time is not known, then it is called Censoring.

Three reasons are
(I) A person is lost to follow-up during the study period.
(II) A person does not experience the event before the study ends.
(III) A person withdraws from the study because of death (if death is not the event of interest) or some other reason (e.g., adverse drug reaction or another competing risk).
These are all examples of right-censoring. (IV) True survival time is equal to or greater than observed survival time.

If the actual survival time is less than or equal to the experimental survival time, then it is defined as left-censored.
If the real survival time is within a known time interval, then it is defined as Interval-censored.
Types of right-censoring occur when a study is designed to end after specific years of follow-up. In this case, everyone who does not have an event observed during the course of the study is censored at particular years.
The study is designed to end after particular years, but censored subjects do not all have the same censoring time. This is the primary type of right-censoring, which we defined as Type I censoring-Random. A study ends when there is a prespecified number of events which is defined as Type II censoring.

1.4 Study Design and Survival Analysis

Suppose the survival function at time t defined as $S(t)$. The probability free time is t. Cumulative incidence function at time t is presented as $F(t) = 1 - S(t)$ as the complementary probability of the event that is occurred time by t. Conventionally, log-rank test is suitable for survival analysis and sometimes assumed with exponential distribution.
However, the power depends on Type I error (i.e., level of significance). Suppose the difference of interest is Δ under H_A.
Now survival analysis can be performed by sequential way or by fixed sample size.
Suppose two groups are created by
Group 1: $(Y_{11}, ...Y_{1n_1})$
Group 0: $(Y_{01}, ...Y_{0n_0})$

$$Y_{1j} \sim N(\mu_1, \sigma^2) \tag{1.14}$$

$$Y_{0j} \sim N(\mu_0, \sigma^2) \tag{1.15}$$

Now the hypothesis is prepared as
$H_0 : \mu_1 = \mu_0 \Rightarrow H_0 : \Delta = 0$ where $\Delta = \mu_1 - \mu_0$

Now the Z-statistics is prepared as

$$Z = \frac{\bar{Y_1} - \bar{Y_0}}{\sqrt{\sigma^2(\frac{1}{n_1} + \frac{1}{n_0})}} \tag{1.16}$$

Let σ^2 is not defined. Conventionally, it can be estimated by s^2. Test statistics can be assumed as $Z \sim N(0,1)$ under the null hypothesis H_0. Similarly, sample sizes are equal for two arms as $n_0 = n_1 = \frac{n}{2}$.

$$Z = \frac{\bar{Y_1} - \bar{Y_0}}{\sqrt{s^2(\frac{1}{n/2} + \frac{1}{n/2})}} = \frac{\bar{Y_1} - \bar{Y_0}}{2s/\sqrt{n}} \tag{1.17}$$

Now the task is to calculate the sample size by defining the critical value c by rejecting the null hypothesis while it is true. Compute the probability of rejecting the null while the alternative is genuine, substituting c from above.

Suppose the α is fixed with probability of rejecting the null hypothesis as

$$\alpha = P(|\bar{Y_1} - \bar{Y_0}| > c|H_0) \tag{1.18}$$

$$\alpha = P(\frac{|\bar{Y_1} - \bar{Y_0}|}{2\sigma/\sqrt{n}} > \frac{c}{2\sigma/\sqrt{n}}|H_0) \tag{1.19}$$

$$\alpha = P(|Z| > \frac{c}{2\sigma/\sqrt{n}}) = 2 * \Phi(-\frac{c}{2\sigma\sqrt{n}}) \tag{1.20}$$

$$z_{1-\alpha/2} = \frac{c}{2\sigma/\sqrt{n}} \tag{1.21}$$

or

$$c = \frac{z_{1-\alpha/2}2\sigma}{\sqrt{n}} \tag{1.22}$$

Let z_γ is presented as $\Phi(z_\gamma) = Pr(Z < z_\gamma) = \gamma$. Compute the probability of rejecting the null when H_A as true. Suppose the probability of a Type II error as

$$\beta = P(\text{accept } H_0|H_\alpha) \tag{1.23}$$

So

$$1 - \beta = P(\text{reject } H_0|H_\alpha) \tag{1.24}$$

$$1 - \beta = P(|\bar{Y_1} - \bar{Y_0}| > c|H_\alpha) \tag{1.25}$$

$$1 - \beta = P(\frac{|\bar{Y_1} - \bar{Y_0}| - \Delta}{2\sigma/\sqrt{n}} > \frac{c - \Delta}{2\sigma/\sqrt{n}}|H_\alpha) \tag{1.26}$$

$$1-\beta = P(Z > \frac{c-\Delta}{2\sigma/\sqrt{n}}) \tag{1.27}$$

so, we get

$$z_\beta = -z_{1-\beta} = \frac{c-\Delta}{2\sigma/\sqrt{n}} \tag{1.28}$$

It can be substitute as c by

$$-z_{1-\beta} = \frac{z_{1-\alpha}2\sigma/\sqrt{n} - \Delta}{2\sigma/\sqrt{n}} \tag{1.29}$$

$$-z_{1-\beta} = z_{1-\alpha/2} - \frac{\Delta}{2\sigma/\sqrt{n}} \tag{1.30}$$

Suppose the equation in terms of sample size of a given power is $1-\beta$, and the level of significance α:

$$z_{1-\alpha/2} + z_{1-\beta} = \frac{\Delta}{2\sigma/\sqrt{n}} \tag{1.31}$$

$$z_{1-\alpha/2} + z_{1-\beta} = \frac{\Delta\sqrt{n}}{2\sigma} \tag{1.32}$$

$$n = \frac{(z_{1-\alpha/2} + z_{1-\beta})^2 4\sigma^2}{\Delta^2} \tag{1.33}$$

If the "effect siz" increased, then the study power increased. Suppose Δ/σ is the number of standard deviations for two means with the equal variances.

1. As n increases, then the power increases.
2. If n is fixed and then σ as the power increases with Δ.
3. If n and Δ fixed then power decreases with σ.
4. It is optimum to assign an equal size to the patients into two groups as $(n_1 = n_0 = n/2)$ by maximizing the power.

1.5 Survival Analysis Objective

The objectives of survival analysis include

(I) The study of patterns of event times. (II) The comparison of distributions of survival times in different groups of individuals. (III) Examining whether and by how much some factors affect the risk of an event of interest.

1.6 Non-Parametric Approach for Survival Analysis

Suppose total number of independent observations are n with

$$(t_1, d_1), (t_2, d_2), ...(t_n, d_n) \tag{1.34}$$

for (T, D). Now the survival duration is $t_1, t_2, ... t_n$. The event is defined as $d_i = 1$, if event occurred. Similarly if the patients alive, then censoring time is presented as $d_i = 0$. Counting process can be presented as

$$Y_i(t) = 1\{t_i \geq t\}, \bar{Y}(t) = \sum_{i=1}^{n} Y_i(t) \tag{1.35}$$

$$N_i(t) = 1\{t_i \leq t, d_i = 1\}, \bar{N}(t) = \sum_{i=1}^{n} N_i(t) \tag{1.36}$$

Risk set is presented as $R(t) = \{i; t_i \geq t\}$. It is assumed that T_{surv} and T_{cons} are independent. Now $S(t)$ and $G(t)$ can be obtained by Kaplan-Meier estimator as

$$\hat{S}_{KM}(t) = \prod_{s \leq t} (1 - \frac{\Delta\bar{N}(s)}{\bar{Y}(s)}) \tag{1.37}$$

Number of events untill time t is presented as $\Delta\bar{N}(t)$. Greenwood's formula presented as

$$se^2(\hat{S}_{KM}(t)) = \hat{S}^2_{KM}(t) \sum_{s \leq t} \frac{\Delta\bar{N}(s)}{\bar{Y}(s)(\bar{Y}(s) - \Delta\bar{N}(s))} \tag{1.38}$$

Pointwise confidence interval with survival function $S(t)$ is presented by standard error as $se(\hat{S}_{KM}(t))$. Simlarly, the confidence interval is $\hat{S}_{KM}(t) \pm z_{1-\alpha/2} se(\hat{S}_{KM}(t))$, with $z_{1-\alpha/2}$ the $1 - \alpha/2$ percentile of the standard normal distribution.

1.7 Log-Rank Test

Let the two groups of individuals are $S_1(t)$ and $S_2(t)$. Null hypothesis is $H_0 : S_1(t) = S_2(t) \forall t$. Contingency table is presented as 2×2 with variance

$$V_i = \frac{R_{1i} R_{2i} C_{1i} C_{2i}}{n_i^2 (n_i - 1)} \tag{1.39}$$

Log-rank test is presented as

$$X^2 = \frac{(O_1 - E_1)^2}{V} \sim \chi_i^2 \tag{1.40}$$

Test statistics can be presented as

$$X^2 = \frac{(O_1 - E_1)^2}{E_1} + \frac{(O_2 - E_2)^2}{E_2} \sim \chi_i^2 \tag{1.41}$$

1.8 Median Follow-Up Time Calculation

Different types of survival data are available. Models useful to understand statistical interpretation and application in the entire field are well accepted. The frequentist approach for right-censored data is well received. However, semiparametric Bayesian methods in survival analysis have become advanced and complex but getting more attention in applied science to their potential fit to the complex data. Now the semiparametric method is generated with parametric and non-parametric sections. The parametric portion is assumed to have a prior distribution with possibly unknown hyperparameters. But the non-parametric part is prepared by a stochastic process.

1.9 Survival Data

The terminal event of cancer treated patient is dead. The survival duration is counted as days from the registration date to the treatment setting until the last follow-up or death as survival duration.

Survival analysis refers to a set of statistical techniques used to investigate the time for an event of interest to take place. Methods of this analysis are used in various fields, such as event-history analysis in sociology and failure-time models in engineering in the name of reliability analysis, and many more. It is also widely used in medical research, specifically cancer studies. A few examples of cancer research where these tools are used are as follows: survival of leukemia patients and their time of remission, their recurrence-free survival, and time to develop any other competing event during treatment of cancer [1].

Traditionally, in survival studies final event of interest used to be merely the death status of patients. Nevertheless, the scope has become broader and more comprehensive nowadays, as it incorporates various events like progression, recurrence, relapse of the disease, or recovery from the illness as endpoints of a study. Like other statistical approaches, the analysis can be performed using both parametric and non-parametric methods. Weibull, exponential, log-logistic, and lognormal are commonly used parametric survival models. In contrast, non-parametric and semi-parametric analyses are primarily performed using the Kaplan-Meier method or Cox proportional hazards (Cox PH) regression modeling, respectively [2]. Kaplan-Meier plot is used to visualize the survival curves, and the log-rank test is used to compare the survival of two or more groups and the. Similarly, the Cox PH model describes covariates' effect, considered simultaneously, on survival time. Cox PH model, the measure of influence is the hazard rate, which is the risk of failure given that the participant has survived up to a specific time. While comparing two independent groups using the Cox PH model and hazard ratios are analogous to the odds ratio of logistic regression.

1.9.1 Multiple event-time data

One cancer-treated patient can be observed with several events. Events could be presented as complete response, partial response, progression disease, and stable disease. But none of them is presented as terminal events. Survival analysis could be helpful in working with multiple events. Perhaps, it considers the events are independent of each other and work as separate event-time models.

1.9.2 Multivariate survival data

Treatment plays an influential role in better survival. But other factors like old age, gender, and genomic profiles play a similar influencing role for better or poor survival. A single survival model can accommodate all influencing factors as a multivariate survival model. Multivariate survival analysis helps us to understand different aspects' combined role in survival outcomes.

1.9.3 Univariate survival models

Survival time is defined by T as a random variable. Now T given a covariate vector x by $S(t|x)$ with hazard function $h(t|x)$. Now Cox's proportional hazard model is presented as

$$h(t|x) = h_0(t)\exp(\beta x) \tag{1.42}$$

The random variable of interest-survival time is denoted by T. We suggest the survivor function of T given a covariate vector x (the vector of explanatory and design variables) by $S(t|x)$ and denote the corresponding hazard function by $h(t|x)$ (when T is absolutely continuous). A popular semiparametric approach for analyzing univariate survival data makes use of Cox's (1972a) proportional hazards model:

$$h(t|x) = h_0(t)\exp(\beta x) \tag{1.43}$$

Now $h_0(.)$ is baseline hazard function. The regression coefficient is defined as β. Conventionally, β is assumed constant over time. Now β is function of time t. It can be considered a time-varying constant. The unknown baseline function is presented as $h_{0(.)}$ as the unknown baseline hazard function. Now the entire interval is presented as $I_1, I_2, ..., I_{g+1}$ with $I_j = [a_{j-1}, a_j]$ for $j = 1, 2, ..., g+1$ by $a_0 =$ with $a_{g+1} = \infty$. Let the survival Cox model presented as

$$S(t|x) = \exp\{-H_0(t)\exp(\beta x)\} \tag{1.44}$$

with β and x is unchaged over time as $H_0(t) = \int_0^t h_0(u)du$ presented as cumulative baseline hazard. So

$$S(a_j|\beta, x) = \prod_{k=1}^{j}(1-\alpha_k)^{\exp(\beta x)} \text{for} j = 1, ..., g \tag{1.45}$$

1.9.4 Multivariate survival models

Semi-parametric multivariate survival model can be classified into two groups marginal and conditional models. The dependent model can be performed by semi-parametric Bayesian and likelihood analysis. In this context, the marginal model is popular in generalized estimating equations methodology. However, longitudinal data is ideal for working with generalized estimating equations. Perhaps, survival data coupled with longitudinal data is suitable for generalized estimating equations methodology.

Now the equation can be presented as

$$h_{li}(t|x_{li}, w_l) = h_0(t)\exp(\beta x_{li})w_l \tag{1.46}$$

Regression coefficient β and the baseline hazard function is presented as h_0 if the frailty random variables are presented as w_l. Suppose $h_0(.)$ is the unknown baseline hazard function about occurrences of the event of a patient. The variance of frailty is not known for w_I. Alternatively, favorable stable distribution also suitable. Different variations can be accommodated for the frailty model.

$$h_{li}(t|x_{li}, w_l) = h_0(t)\exp(\beta x_{li})w_{li} \tag{1.47}$$

The term w_{li} is stochastically dependent for patients but independent from one patient to another. This model can accommodate the negative correlation matrix for the w_{li} in the patient.

1.9.5 Doubly interval-censored survival data

Interval-censored data is an essential feature in survival data mechanisms. Many times, we do overlook the presence of interval censoring. Commonly, the survival duration can be split into different intervals and thus help to accommodate the survival analysis. The model is formulated as

$$P(T_2 \in I_k|x) = f_k(x) = (p_1, \ldots.., p_{k-1})^{e^{x\beta}}(1 - p_k^{x\beta}) \tag{1.48}$$

Further, survival duration is spread as T_1 and T_2 for the independent covariates.

1.9.6 Frequentist approach

A conventional approach to working with survival data is suitable for working with right-censored data. It works with a time-dependent covariate but assumes that the time-varying covariate is constant over time. The frequentist method cannot verify the model assumption such as proportional hazards, constancy of the covariate effect over time, etc.

1.10 Bayesian Prior Assumptions for Survival Analysis

Bayesian Inference generated with posterior probability of the model m^* in M with Bayes theorem as

$$p(m^*|y) = \frac{p(y|m^*)p(m^*)}{\sum_{m \in M} p(y|m)p(m)} \tag{1.49}$$

and

$$p(y|m) = \int_{\Theta_m} p(y|\theta_m, m)\pi(\theta_m)d\theta_m \tag{1.50}$$

Suppose the model m belongs to the parameter space Θ_m.Now the marginal distribution $p(y|m)$ of the model m may leads to "marginal distribution." The primary factor Bayes Factor provides the data in favor of one model with respect to another prior model probability with marginal $p(y|m)$ data model if the prior $\pi(.)$ is not a proper density. Now the Bayes Factor (BF) is defined the models by M_1 and M_2. Thus $p(M_2) = 1 - p_(M_2)$. The prior odds becomes in favor of M_1 as $p(M_1)/p(M_2)\}$. Similarly, the posterior odds becomes

$$p(M_1|y)/p(M_2\ y) = p(y|M_1)p(M_1)/\{p(y|M_2)p(M_2\}. \tag{1.51}$$

Bayes Factor defines the odds ratio as

$$BF_{1,2} = \frac{\text{Posterior odds of } M_1}{\text{Prior odds of } M_1} \tag{1.52}$$

$$BF_{1,2} = \frac{p(y|M_1)}{p(y|M_2)} \tag{1.53}$$

The Bayes Factor defines as

$$BF_{2,1} = 1/BF_{1,2} \tag{1.54}$$

Now M_1 can contain two models, and the ratio of the data marginal of two models generates the BF through the model space. It depends on the prior odds of a particular probability of the individual models to the marginal data.

1.10.1 Prior in survival analysis

The progress of Bayesian in survival started after immediate work after Cox [3]. Sometimes, non-parametric Bayesian pieces with a prior random process. However, it generates a different distribution function. Now the non-parametric statistical portion can work separately with Bayesian modeling. Accommodating additional covariates makes it a little complex.

1.10.2 Dirichlet process prior

The Dirichlet process is presented for random probability. It works suitably with prior distribution and generates distribution function $F(.)$. There is extensive literature on the Dirichlet process prior, and it is essential to understand about Dirichlet process prior. Perhaps, it is difficult to work with the Dirichlet process prior in terms of the hazard function.

1.11 Illustration Using R

The results of the analysis are presented with different figures. Figure 1.1. shows the Kaplan-Meier figure using R. It is presented with two curves, and both curves make the treatment-wise comparison of survival. The Kaplan-Meier plot presented with a shaded scale and confidence interval is in figure 1.2. The estimates of the forest plot are presented in figure 1.3. We can also prepare the Kaplan-Meier curve by using the library "survminer" using R and given in figure 1.4. This similar graph is extended with a confidence interval with a similar package in figure 1.5. It is possible to make the figure more informative by adding a risk table illustrated in figure 1.6. The cumulative plot with "survminer" package is presented in different ways in figure 1.7 and figure 1.8.

Call survival library using R

```
library(survival)
data(pbc)
pbc
```

Call survival library using R

```
library(survival)
fit1=survfit(Surv(time, status) ~ 1, data=pbc)
summary(fit1)
```

Call survival library using R

time	n.risk	n.event	survival	std.err	lower 95% CI	upper 95% CI
41	186	2	0.98	0.00	0.97	1.00
43	184	1	0.98	0.00	0.96	1.00
51	183	1	0.97	0.01	0.95	0.99
71	182	1	0.97	0.01	0.95	0.99
77	181	1	0.96	0.01	0.94	0.99
94	180	1	0.96	0.01	0.93	0.99
110	179	1	0.95	0.01	0.92	0.98
111	178	1	0.95	0.01	0.92	0.98
130	177	1	0.94	0.01	0.91	0.97
131	176	1	0.94	0.01	0.90	0.97

Survival plot using R

```
fit1=survfit(Surv(time, status) ~ trt, data=pbc)
autoplot(fit1)
```

Analysis with high dimensional data

```
library(SurvHiDim)
data(hnscc)
hnscc
```

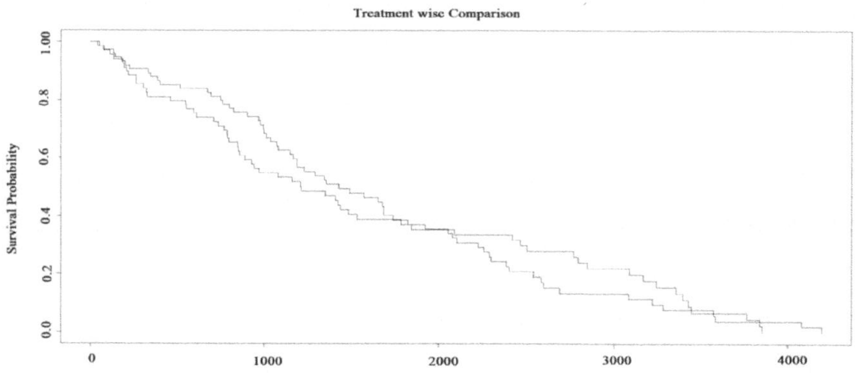

FIGURE 1.1: Kaplan-Meier plot using R.

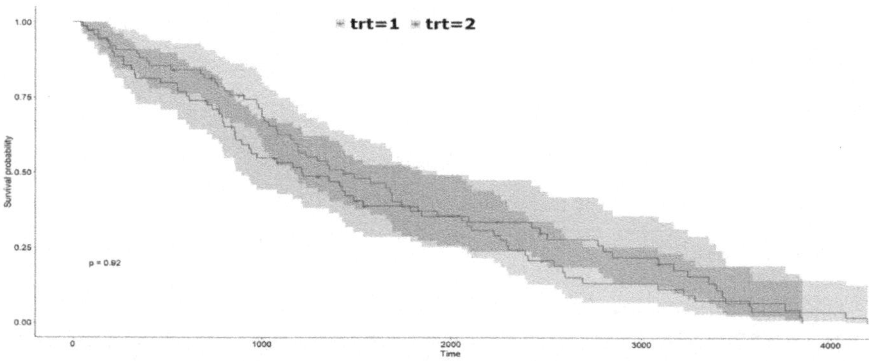

FIGURE 1.2: Kaplan-Meier plot using R.

Survival analysis with high dimensional data

library(SurvHiDim)

survdesc(m,n,survdur,event,data)

m=Starting column number form where study variables of high dimensional big data will get selected.

n=Ending column number till where study variables of high dimensional data will get selected.

survdur=Column name of survival duration event, a string value. i.e. "os".

event=Column name of survival event, a string value. i.e "death".

data=High dimensional data containing the survival, progression and several covariates observations.

Function to call high dimensional survival data

```
library(SurvHiDim)
fit1=survdesc(m=10,n=20,survdur="os",event="death",data=hnscc)
fit1
```

Outcome of the survdesc

Variables	HR	LCL	UCL	p-value
SLC4A4	1.02	0.96	1.09	0.449
HNF1B	1.07	0.92	1.24	0.340
TOX3	1.02	0.97	1.08	0.313
CLDN3	1.00	0.96	0.04	0.986
RGL3	1.00	0.91	1.09	0.971
LMO3	1.02	0.96	1.09	0.371
HGD	1.04	0.94	1.14	0.427
C7	1.01	0.97	1.06	0.410
LRP2	1.09	1.00	1.19	0.036
MYEF2	0.97	0.90	1.05	0.556
C2orf72	1.07	1.00	1.14	0.049

Function to call high dimensional survival data

```
library(SurvHiDim)
fit1=survdesc(m=10,n=20,survdur="os",event="death",data=hnscc)
fit1
```


Conventional Cox PH Multivariate model

```
library(survival)
model=coxph( Surv(os,death) ~ GJB1+PPP1R9A+HPN+SLC4A4+
HNF1B+TOX3+CLDN3+RGL3+LMO3+HGD+C7
+ LRP2+MYEF2+C2orf72+GOLT1A+C16orf89+DCDC2
+PLEKHB1+LONRF2+GAL3ST1,data=hnscc)
```

Listing 1.3: Calculating median duration of survival using R

```
The overall median survival is calculated as 310 days.
library(survival)
data(cancer)
fit1<-survfit(Surv(time,status) ~ 1, data=cancer)
print(fit1)
Call:
survfit(formula=Surv(time, status)~1,data=cancer)

        n events median 0.95LCL 0.95UCL
[1,] 228    165    310     285     363
```

Listing 1.4: Calculating group wise median duration of survival using R

```
The gender wise median survival is calculated
as 270 and 426 days.
library(survival)
data(cancer)
fit2 <- survfit(Surv(time,status) ~ sex,data=cancer)
print(fit2)
Call: survfit(formula=Surv(time,status)~sex,data=cancer)
         n events median 0.95LCL 0.95UCL
sex=1 138    112    270     212     310
sex=2  90     53    426     348     550
```

Listing 1.5: Preparation of survival plot with survminer

```
library(survival)
library(survminer)
data(pbc)
fit1=survfit(Surv(time, status)~ sex , data=cancer)
ggsurvplot(fit1 , linetype = ``strata", pval = TRUE,
                    palette = ``grey")
```

Listing 1.6: Survival plot with confidence interval and P-value

```
library(survival)
library(survminer)
data(pbc)
fit1=survfit(Surv(time, status)~ sex , data=cancer)
ggsurvplot(fit1 , linetype = ``strata",conf.int = TRUE,
pval = TRUE,palette = ``grey")
```

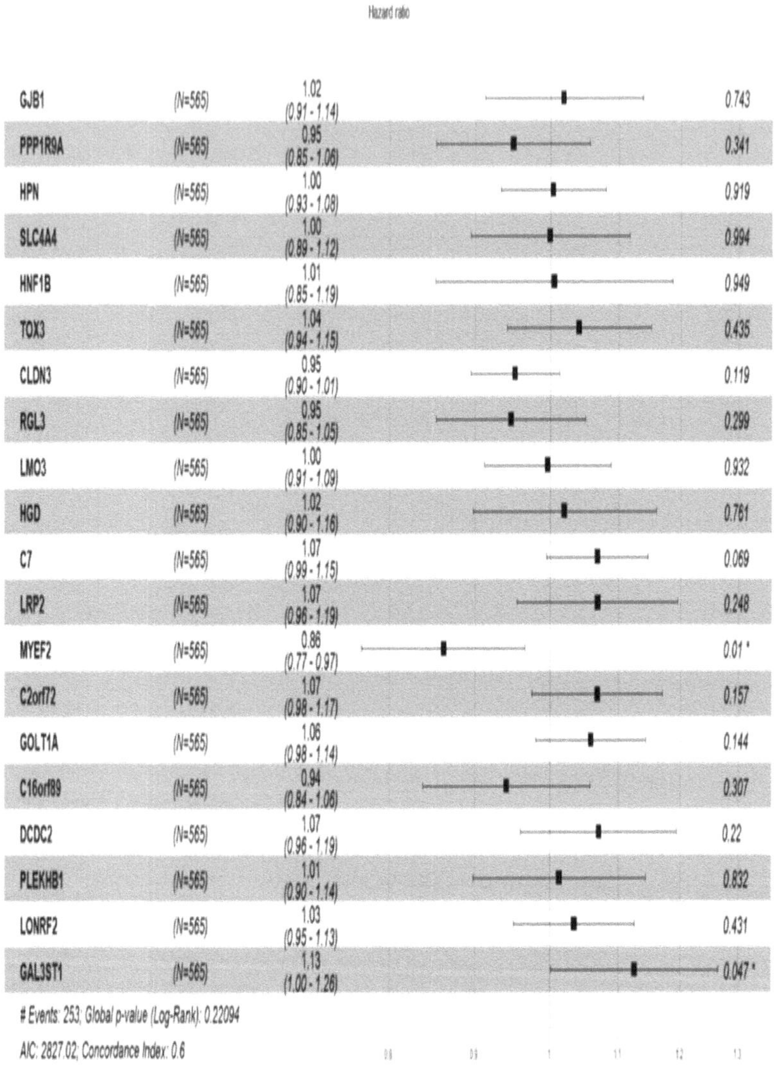

FIGURE 1.3: Results obtained on Conventional Cox PH Model.

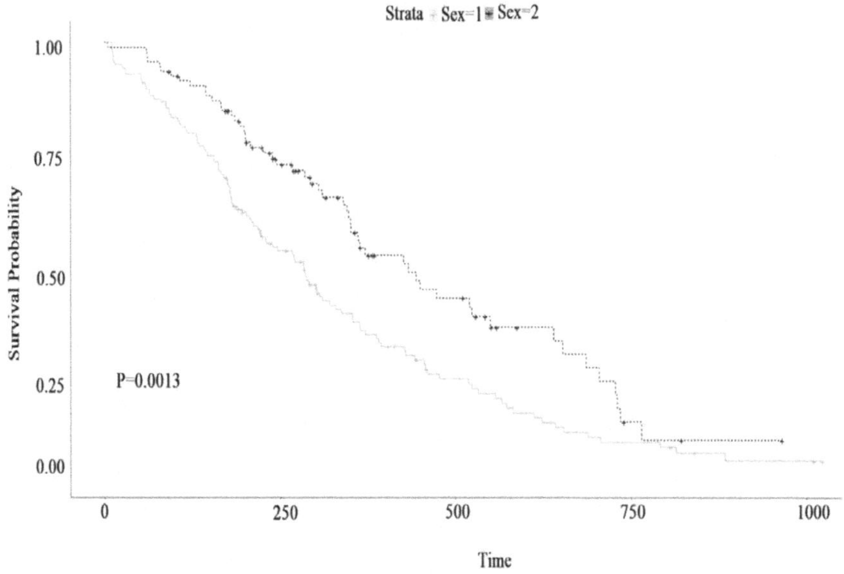

FIGURE 1.4: Preparation of survival plot with survminer.

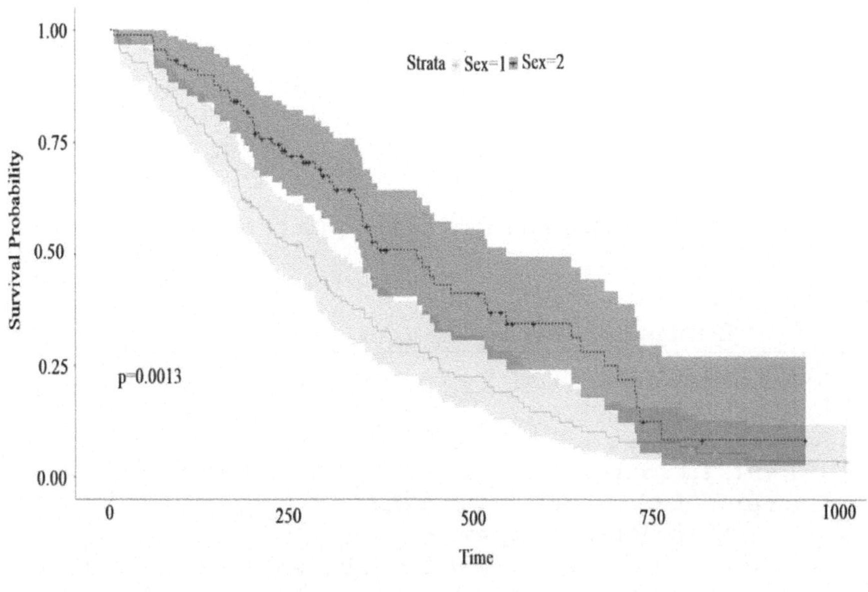

FIGURE 1.5: Survival plot with confidence interval and P-value.

Listing 1.7: Survival plot with risk table

```
library(survival)
library(survminer)
data(pbc)
fit1=survfit(Surv(time, status)~ sex , data=cancer)
ggsurvplot(fit1, linetype = ``strata",conf.int = TRUE,
risk.table = TRUE, pval = TRUE,palette = ``grey")
```

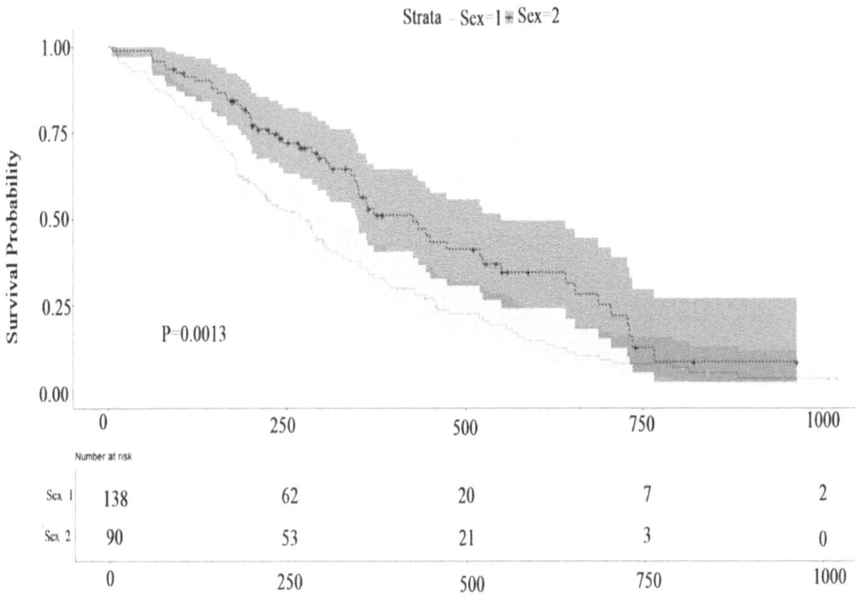

FIGURE 1.6: Survival plot with risk table.

Listing 1.8: Plot cumulative event in teh survival plot

```
library(survival)
library(survminer)
data(pbc)
fit1=survfit(Surv(time, status)~ sex , data=cancer)
ggsurvplot(fit1, conf.int = TRUE,
           palette =``grey",
           risk.table = TRUE, risk.table.col = ``strata",
           fun = ``event")
```

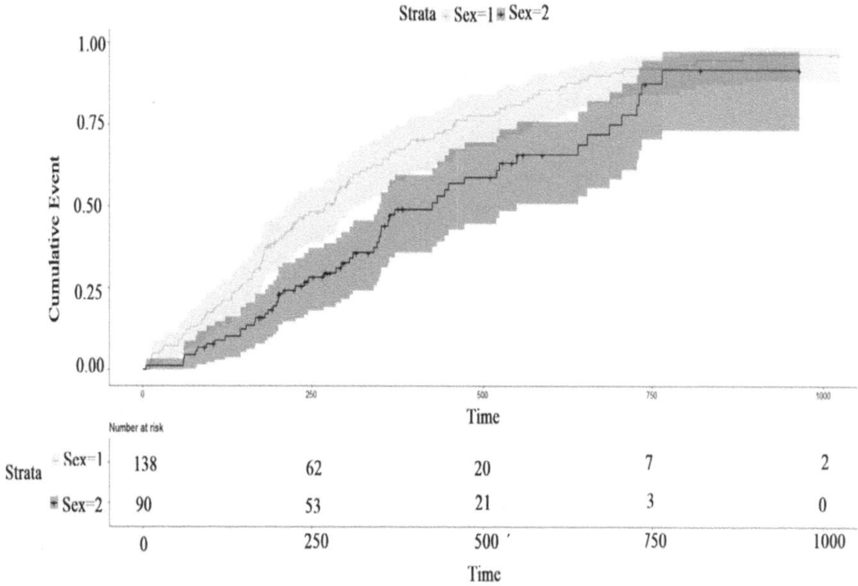

FIGURE 1.7: Plot cumulative event in teh survival plot.

Listing 1.9: Inclusion of cumulative hazard function in the survival plot

```
library(survival)
library(survminer)
data(pbc)
fit1=survfit(Surv(time, status)~ sex , data=cancer)
ggsurvplot(fit1 , conf.int = TRUE,
           palette =``grey",
           risk.table = TRUE, risk.table.col = ``strata",
           fun = ``cumhaz")
```

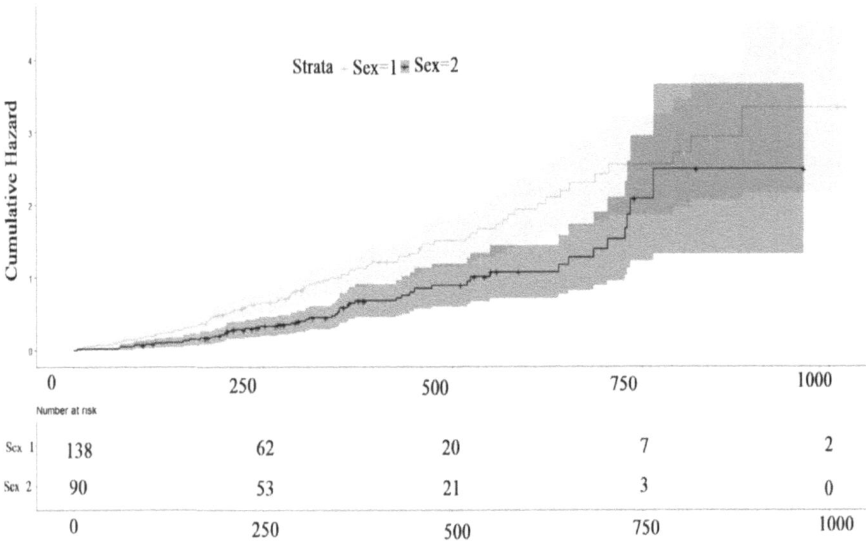

FIGURE 1.8: Inclusion of cumulative hazard function in the survival plot.

Chapter 2

Cox Proportional Survival Analysis

2.1 Introduction

The survival analysis is based on the dependent of survival distribution. The survival function is presented with the probability distribution up to a point t.

$$S(t) = Pr(T > t), 0 < t < \infty \tag{2.1}$$

The range of the function is between 1 and 0. It can be decreased over time and dropped below 0. Survival function is often presented as hazard function. The hazard function is the instantaneous failure rate. Now the probability of hazard function is presented as

$$h(t) = lim_{\delta \to 0} \frac{Pr(t < T < t + \delta | T > t)}{\delta} \tag{2.2}$$

This function is also known as the intensity function or the force of mortality.

2.2 Cox Proportional Hazard

Suppose $(T_1, T_2, ...T_n)$ is defined as Life times and, $(C_1, C_2, ...C_n)$ as the Censoring times. Let it is defined as $Y_i = min(T_i \leq, c_i), i = 1, 2, ...n$. Now our data is $(Y_i, \delta_i), i = 1, 2, ...n$. Suppose corresponding to Y_i we have covariates $\underline{X}_i = (X_{1i}, X_{2i}, ...X_{pi})'$. Because of the censoring, it is difficult to deal with a linear regression of Y_i on $\underline{X}_i$. It is easier to relate $\underline{X}_i$ to the hazard function $\lambda(t)$. Let $\lambda(t|\underline{x})$ denote the hazard function of an individual with covariate $\underline{X}$. Suppose $\lambda(t|\underline{x}) = \lambda_0(t)g(\underline{x})$ is the multiplication hazard model, where

DOI: 10.1201/9781003185598-2

$\lambda_0(t)$ = function of t but not of $\underline{x}$(baseline hazard). $g(\underline{x})$ = function of $\underline{x}$ but not t except for covariates dependent on t. Note that

$$\frac{h(t|\underline{x}_1)}{h(t|\underline{x}_2)} = \frac{g(\underline{x}_1)}{g(\underline{x}_2)} \tag{2.3}$$

indepdendent of t.
i.e.,

$$h(t) = h_0(t)e^{\underline{x}\beta} \tag{2.4}$$

Note that $h_0(t)$ and $g(\underline{x})$ must be positive and hence a choice of $e^{\underline{x}'\beta}$ for $g(\underline{x})$ is justified. Here

$$\frac{h(t|\underline{x}_1)}{h(t|\underline{x}_2)} = e^{(x_1' - x_2')\beta} \tag{2.5}$$

The Cox model specifies the hazard for individual i as

$$h_i(t) = h_0(t)e^{X_i(t)\beta} \tag{2.6}$$

where h_0 is an unspecified positive function of time called the baseline hazard, and β is a $p \times 1$ column vector of coefficients.

2.2.1 Hazard ratio

The hazard ratio for two subjects with fixed covariate vectors X_i and X_j are

$$\frac{h_i(t)}{h_j(t)} = \frac{h_0(t)e^{X_i\beta}}{h_0(t)e^{X_j\beta}} = \frac{e^{X_i\beta}}{e^{X_j\beta}} \tag{2.7}$$

The term

$$\frac{h_i(t)}{h_j(t)}$$

is constant over time then the model is known as the proportional hazards model.

2.2.2 Partial likelihood function

Let $y_{(1)}, ..., y_{(n)}$ be the ordered observation time for each δ_i = 1(uncensored time point). Let R_j be the risk-set at time $(y_{(j)}, \delta_j)$. and $\underline{x}_j$

the covariate corresponding to $y_{(i)}$.

$$P[\text{a death in}[y_{(j)}, y_{(j)} + \Delta y|R_j] = \sum_{k \epsilon R_j} e^{\underline{x}_k^k \beta} \lambda_0(y_{(j)})\Delta_y \tag{2.8}$$

$$P[y_{(j)}|\text{one death at time } y_{(j)}] = \frac{e^{\underline{x}_j^{'} \beta}}{\sum_{k \epsilon R_j} e^{\underline{x}_j^{'} \beta}} \tag{2.9}$$

The maximum partial likelihood estimator is found by solving the partial likelihood equation. Now $\hat{\beta}$ is consistent and asymptotically normally distributed with mean β with variance $\{EI(\beta)\}^{-1}$. Let that subjects $i = 1, ..., n_1$are in stratum 1, subjects $n_{l+1}, ..., n_l + n_2$ are in stratum 2, and so on. The hazard for an individual i, who belongs to stratum k is

$$\lambda_k(t)e^{X_i \beta} \tag{2.10}$$

The overall loglikelihood

$$l(\beta) = \sum_{k=1} l_k(\beta) \tag{2.11}$$

The score vector and information matrix are $U(\beta) = \sum U_k(\beta)$ and $I(\beta) = \sum I_k(\beta)$ respectively.

2.2.3 Wald score and Likelihood ratio tests

The Wald score, and likelihood ratio, are helpful for the Cox partial likelihood to test hypotheses about β. The likelihood ratio test is $2(l(\hat{\beta}) - l(\beta^0)$, twice the difference in the log partial likelihood at the initial and final estimates of $\hat{\beta}$.

2.3 Cox Proportional Diagnostics Test

Cox Model is defined as

$$h(t) = h_0(t)e^{x^{'} \beta} \tag{2.12}$$

Now x presented with fixed time covariates and

$$F(t) = 1 - S(t) \sim R(0,1) \tag{2.13}$$

It is presented as $S(t) \sim R(0,1)$ Now

$$S(t) = e^{-\Delta(t)} \tag{2.14}$$

$$\frac{dS(t)}{d\Delta(t)} = -e^{-\Delta(t)} \tag{2.15}$$

The probability density function is presented as

$\Delta(t)$ with $f^*(\Delta(t)) = e^{-\Delta(t)}, 0 < \Delta(t) < \infty$. It is assumed that $\Delta(t)$ with an exp distribution with hazard rate 1.

Now $\bar{\beta}$ can be estimated as β. The Breslow's estimation for the baseline hazard function is presented as $h_0(y_i)$ with

$$\hat{h}_0(y_i) = \frac{1}{\sum_{j \epsilon R_i} e^{x'_j \beta}} \tag{2.16}$$

The cumulative baseline hazard function $h_0(t)$ is

$$\hat{\Delta}(t) = \sum_{y_i \leq t} \frac{1}{\sum_{j \epsilon R_i} e^{x'_j \beta}} \tag{2.17}$$

2.3.1 Cox-snell residual

Once the Cox PH model is fitted, then there is a certain chance of having the error. This error is presented by the Cox-Snell residual.

$$r(y_j) r_j = \hat{\Delta}_0 e^{x'_j \hat{\beta}} \tag{2.18}$$

If the model is true then the $r'_j s$ should sample from a unit exponential distribution. The Nelson-Aalen estimation of the cumulative hazard rate of the $r'_j s$ can be looked at

$$\hat{\Delta}(t) = \sum_{y_i \leq t} \frac{d_i}{n_i} \tag{2.19}$$

It can be compared with the hazard rate of the exponential unit distribution $\Delta^*(t) = t$. Thus plot of $\hat{\Delta}_r(r_j)$ against r_j should be a straight line.

2.3.2 Martingale residual

$$N_j(t) = 1\text{if the ith individual experience the event by time t} \tag{2.20}$$
$$= 0\text{otherwise} \tag{2.21}$$

$$I_j(t) = 1\text{if the individual is understudy at t} \tag{2.22}$$
$$= 0\text{otherwise} \tag{2.23}$$

Residual $\hat{M}_j = N_j(\infty) - \int_0^{\infty} I_j(t) e^{x_j \hat{\beta}} d\hat{\Delta}_i(t)$. For right censored data,

$$\hat{M}_j = \delta_j - \hat{\Delta}_0(y_j) e^{x_j \hat{\beta}} = \delta_j - r_j \tag{2.24}$$

$$\sum \hat{M}_j = 0 \tag{2.25}$$

$$h(t) = h_0(t) e^{\beta_i x_i} e^{\sum_{i=1}^{k} \beta_i x_i} \tag{2.26}$$

and fit a model based on $x_2, ...x_p$.

2.4 Mean and Median Survival Time

The expected value of the survival time is defined as the mean survival time as,

$$\mu = E(T) = \int_0^{\infty} t f(t) dt \tag{2.27}$$

Using standard integration by parts and using the fact that $f(t) = -d/dtS(t)$ may be written as

$$\mu = \int_0^{\infty} S(t) dt \tag{2.28}$$

It can be written as $f(t) = -d/dtS(t)$, while all subjects get failed. But it is not always practical in cancer research. It may make the area under the curve infinite. Theoretically, the survival curve is not possible to generate by

Kaplan-Meier survival while the curve does not merge to zero. The median survival time is presented by time t as $S(\infty) = 1/2$.

2.5 Stratified Cox Proportional Hazard Test

The stratified Cox proportional hazard analysis can be performed in two ways. One approach is by adjusting the covariates, and another one is the inclusion of multiple covariates as regression terms for the hazard function. Suppose the covariates are adjusted for the small number of categories with level G, we may create a stratified log-rank test. Present the null hypothesis as $H_0 : h_{oj}(t) = h_{1j}(t)$ for $j = 1, 2, ..., G$. Similarly, we can compute the test statistics U_{0g} and variance V_{0g}, with $g = 1, ..., G$ is the group indicator. The test statistic can be presented as

$$\chi^2 = \frac{(\sum_{g=1}^{G} U_{0g})^2}{\sum_{g=1}^{G} V_{0g}^2} \tag{2.29}$$

Fundamentally, the stratified Cox proportional hazard is the application of chi-square test statistics, better known as the log-rank test.

2.6 Schoenfeld Residuals

Residuals from a fitted model are most easily understood when they express in some way the difference between observed data values and the corresponding fitted (predicted) values.

A large amount of data available for each chosen x is significant to making a decision about the plot. At the time "i" of death or failure of the unit or individual i, there is a risk set R_i; consisting of the individuals who have not yet failed (or been censored) so were at risk of failure at this time. The conditional probability that "i" died, given that there was a failure at t_l is given by

$$p_i = \frac{h_i(t_i)}{\sum_{j \epsilon R_i} h_j(t_i)} = \frac{e^{\beta' x}}{\sum_{j \epsilon R_i} e^{\beta' x_j}} \tag{2.30}$$

The expected value $E(X|R_i)$ is

$$E(X|R_i) = \sum_{k \epsilon R_i} x_k p_k = \frac{\sum_{k \epsilon R_i} x_k e^{\beta' x_k}}{\sum_{j \epsilon R_i} e^{\beta' x_j}} \tag{2.31}$$

The vector is defined as

$$r_i = x_i - E(X|R_i) \tag{2.32}$$

The estimate of schoenfeld residual is

$$\hat{r}_i = x_i - \hat{E}(X|R_i) \tag{2.33}$$

The Schoenfeld residual is closely related to the estimation procedure of Cox's PH model.

2.7 Extended Cox Regression Model

In time-dependent covariate is, $X_b = X_b.g_b(t)$, where X_b is a covariate that has no time dependency. However, because it does not meet the assumption of Proportional Hazard, X_b must have interacted with the time function. If all p covariate does not meet the proportional hazard assumption, then the number of p covariates must interact with time. The time function for covariate-b is defined as $g_b(t)$, obtained by the form of extended cox as follows:

$$h(t, X(t)) = h_0(t)\exp[\sum_{b=1}^{p} \beta_b X_b + \sum_{b=1}^{p} \delta_b X_b g_b(t)] \tag{2.34}$$

where available p_1 covariate that meets the proportional hazard assumptions and exists p_2 which does not meet the proportional hazard assumption with, then obtained the model as follows

$$h(t, X(t)) = h_0(t)\exp[\sum_{a=1}^{p_1} \beta_a X_a + \sum_{b=p_1+1}^{p_2} \beta_b X_b + \sum_{b=p_1+1}^{p_2 \delta_b X_b g_b(t)}] \tag{2.35}$$

with time function $g_b(t)$ which can be used is: $g_b(t) = 0, g_b = t, g_b(t) = Int$

2.8 Illustration Using R

2.8.1 Univariate Cox proportional hazard in high dimensional data

Univariate Cox proportional hazard model is relatively easy to perform with conventional statistical packages. However, it becomes tedious when we must serve it with several thousand variables. It is not feasible to perform a univariate Cox PH analysis with several thousand variables. It becomes challenging to figure out essential variables by looking at model selection criteria among several thousand parameters in high dimensional data. We prepared a single function in R, and it is the SurvHidim package named survdesc. This "survdesc" function starts with m as the number of beginning columns selected from the high-dimensional data set. It takes the value as "n" as the number of end parameters until the columns are chosen at a stretch. The surdur option is there to specify the overall survival parameter for the analysis. Similarly, the event parameter has the provision to define death and AIC to order the variable with a minimum AIC value. If we choose AIC as TRUE, then ascending order will be created; otherwise, not. Estimates of the survival analysis with survdesc function without AIC presented in Table 2.1. Result of Univaraite Cox PH Analysis Results obtained by survdesc function without AIC presented in Table 2.2. Result of multivariate survival analysis in high dimensional data shown in Table 2.3.Result of multivariate survival analysis by specifying variables in high dimensional data presented in Table 2.4.Estimates obtained on Cox Proportional Hazards with MCMC detailed in Table 2.5.

Code of Univaraite Cox PH Analysis Results obtained by survdesc function with AIC ordered for high dimensional data.

```
library(SurvHiDim)
data(hnscc)
survdesc(m=10,n=20,survdur="os",event="death",aic=TRUE,
data=hnscc)
```

TABLE 2.1: Result of univaraite Cox PH analysis results obtained by survdesc function without AIC ordered for high dimensional data.

Variables	HR	LCL	UCL	p-value	AIC
ZMAT1	0.89	0.84	0.96	0.00	2803.65
GAL3ST1	1.13	1.03	1.24	0.00	2806.87
LRP2	1.09	1.00	1.19	0.03	2809.40
C2orf72	1.07	1.00	1.14	0.04	2809.77
XIST	1.02	0.99	1.05	0.05	2810.08
ADH1B	1.03	0.99	1.07	0.10	2810.92
FAM47E	1.06	0.98	1.15	0.10	2810.96
NME5	1.07	0.98	1.16	0.10	2811.09
KCNJ16	1.04	0.98	1.11	0.10	2811.10
CACNA1D	0.94	0.87	1.02	0.15	2811.50
BEX1	1.05	0.98	1.12	0.14	2811.55
GOLT1A	1.04	0.98	1.11	0.15	2811.55
SULT1C2	0.95	0.87	1.03	0.22	2811.98
LONRF2	1.02	0.97	1.08	0.28	2812.38

Code of Univaraite Cox PH Analysis Results obtained by survdesc function without AIC ordered for high dimensional data

```
survdesc(m=10,n=20,survdur="os",event="death",aic=FALSE,
data=hnscc)
```

TABLE 2.2: Result of univaraite Cox PH analysis results obtained by survdesc function without AIC ordered for high dimensional data.

Variables	HR	LCL	UCL	Pvalue	AIC
SLC4A4	1.02	0.96	1.09	0.44	2812.96
HNF1B	1.07	0.92	1.24	0.34	2812.66
TOX3	1.08	0.97	1.08	0.31	2812.53
CLDN3	1.00	0.96	1.04	0.98	2813.53
RGL3	1.00	0.91	1.09	0.97	2813.52
LMO3	1.02	0.96	1.09	0.37	2812.74
HGD	1.04	0.94	1.14	0.42	2812.91
C7	1.01	0.97	1.06	0.41	2812.86
LRP2	1.09	1.00	1.19	0.03	2809.40
MYEF2	0.97	0.90	1.05	0.55	2813.18
C2orf72	1.07	1.00	1.14	0.04	2809.77

Code of multivariate survival analysis in high dimensional data.

```
multicoxa(m=15,n=18,OS="os",event="death",data=hnscc)
```

TABLE 2.3: Result of multivariate survival analysis in high dimensional data.

Variables	HR	LCL	UCL	Pvalue	AIC
LMO3	1.01	0.93	1.09	0.85	2815.08
HGD	1.01	0.91	1.13	0.8	2815.08
C7	1.01	0.96	1.06	0.81	2815.08
LRP2	1.09	0.99	1.19	0.07	2815.08

Code of multivariate survival analysis by specifying variables in high dimensional data.

```
library(SurvHiDim)
multicoxb(C1="GJB1",C2=NULL,C3="HPN",C4=NULL,C5=NULL,
OS="os",event="death",data=hnscc)
```

TABLE 2.4: Result of multivariate survival analysis by specifying variables in high dimensional data.

Variables	HR	LCL	UCL	Pvalue	AIC
GJB1	1.01	0.94	1.08	0.8	2815.07
HPN	1.01	0.95	1.08	0.6	2815.07

SurvHiDim

```
library(SurvHiDim)
data(hnscc)
HiDimLasso(7,105,OS="os",Death="death",hnscc)
```

Define data

```
library(SurviMChd)
data(hnscc)
```

High dimensional data

```
data(hnscc)
nrow(hnscc) [1] 565
ncol(hnscc) [1] 105
```

Survival analysis with interval censoring data

```
library(SurviMChd)
data(hnscc)
survintMC(7,11,Leftcensor="leftcensoring",OS="os",
Death="death",6,hnscc)
```

Survival analysis using Cox Proportional Hazards with MCMC

```
data(mcsurv)
nrow(mcsurv) [1] 42
ncol(mcsurv) [1] 8
survMC(m=4,n=8,duration="OS",event="Death",chains=2,
iter=10,data=mcsurv)
```

The survival function is presented as $y_i = (t_i, d_i)$ with survival outcome. Now the censoring indicator is specified as d_i, and it takes one of the observed survival times t_i for the individual i is uncensored. If $d_i =$ is 0 then it is censored, while true survival time is assumed by T_i. In the era of big data, several predictor variables can be accommodated for survival analysis-predictor variables like genomic and clinical covariates.

TABLE 2.5: Result obtained on Cox Proportional Hazards with MCMC.

Parameter	Var	DIC	Mean(SD)	(2.5%, 97.5%)	Rhat	n.eff
HR	Cov_1	243.9	0.99(0.05)	(1.01,1.07)	3.66	2
Deviance	Cov_1	243.9	226.72(6.32)	(229.52,237.76)	1.69	4
HR.1	Cov_2	244.6	1.01(0.03)	(1.01,1.07)	2.67	3
Deviance.1	Cov_2	244.6	227.40(5.32)	(229.79,236.76)	1.22	8
HR.2	Cov_3	245.2	1.05(0.21)	(1.11,1.40)	3.43	2
Deviance.2	Cov_3	245.2	227.55(5.45)	(230.24,237.28)	1.15	10
HR.3	Cov_4	249.4	1.00(0.01)	(1.02,1.02v	5.08	2
Deviance.3	Cov_4	249.4	230.04(6.85)	(230.92,241.44)	1.13	10
HR.4	Cov_5	243.7	1.46(0.80)	(1.34,3.12)	2.33	3
Deviance.4	Cov_5	243.7	224.17(6.28)	(229.24,232.58)	1.44	5

The Cox proportional model is presented with hazard function with survival time T as the form

$$h(t|X) = h_0(t)\exp(X\beta) \tag{2.36}$$

Now the baseline hazard function is $h_0(t)$ as unspecified. The explanatory variable is X, and the coefficient is β. Now the linear predictor is used to prepare the prognostic index as $X\beta$. Now the Cox model can be estimated by β by maximizing the partial log-likelihood.

$$pl(\beta) = \sum_{i=1}^{n} d_i \log(\exp(x_i\beta)/\sum_{i' \in R(ti)} \exp(X_{i'}\beta)) \tag{2.37}$$

Suppose the risk set is presented as $R(t_i)$ at time t_i. Now the partial log-likelihood can be approximated as Breslow methods. The standard process is by maximizing the partial log-likelihood by Newton-Raphson algorithm. The classical model is not suitable for survival analysis for high-dimensional and correlated data. Now the appropriate approach is to work with Bayesian hierarchical modeling by penalized methods.

$$\beta_j|\gamma_j, s_0 s_1 \sim de(0, (1-\gamma_j)s_0 + \gamma_j s_1) = \frac{1}{(1-\gamma_j)s_0 + \gamma_j s_1}\exp(-\frac{|\beta|}{(1-\gamma_j)s_0 + \gamma_j s_1}) \tag{2.38}$$

The terms s_0 and s_1 are used for scale parameters as strong and weak shrinkage on β_j. The indicator variable is presented as $\gamma_j = 1$ or 0. It can be expressed as

$(1-\gamma_j)de(0, s_0) + \gamma_j de(0, s_1)$. It comes as mixture of shrinkage prior $de(0, s_0)$ with weakly informative prior as $de(0, s_1)$. It presented as spike and slab components of the prior distribution. Suppose the Berllouli distribution is used to prepare the group structure as indicator variable by θ_g:

$$\gamma_j|\theta_g \sim Bin(\gamma_j|1, \theta_g) = \theta_g^{\gamma_j}(1-\theta_g)^{1-\gamma_j} \tag{2.39}$$

If g accommodates with important parameters then it becomes large by θ_g. We can consider the beta prior by $\theta_g \sim beta(a, b)$, setting $a = b = 1$. It generates uniform hyper prior $\theta_g \sim U(0, 1)$ distribution.

2.8.2 Expectation-maximization algorithm

The expectation-maximization(EM) algorithm can be developed to fit the spike-and-slab in Cox models through the estimation of posterior modes of parameters. Adoption of EM algorithm makes the high dimensional data analysis fast and efficient with Cox model. The EM algorithm can be presented as $\theta = (\beta, \gamma, \theta)$:

$$\begin{aligned} logp(\beta, \gamma, \theta|t, d) \propto logp(t, d|\beta, h_0) + \sum_{j=1}^{J} logp(\beta_j|S_j) \\ + \sum_{j=1}^{J} logp(\gamma_j|\theta_g) + \sum_{g=1}^{G} logp(\theta_g) \end{aligned} \tag{2.40}$$

Now $pl(\beta)$ is the partial likelihood described and $S_j = (1-\gamma_j)s_0 + \gamma_j s_1$. The indicator variables γ_j is treated as missing value. Parameters (β, θ) can be estimated by averaging the missing values over the posterior distribution. The E-step of the expectation can be obtained by joint posterior density with respect to conditional posterior distribution of the missing data. Now the conditional expectation of the indicator variable γ_j can be derived as

$$\begin{aligned} p_j^g &= p(\gamma_j = 1|\beta_j, \theta_g, t, d) \\ &= \frac{p(\beta_j|\gamma_j = 1, s_1)p(\gamma_j = 1|\theta_g)}{p(\beta_j|\gamma_j = 0, s_0, s_0)p(\gamma_j = 0|\theta_g) + p(\beta_j|\gamma_j = 1, s_1)p(\gamma_1 = 1|\theta_g)} \end{aligned} \tag{2.41}$$

where $p(\gamma_j = 1|\theta_g) = \theta_g, p(\gamma_j = 0|\theta_g)) = 1-\theta_g, p(\beta_j|\gamma_j = 1, s_1) = de(\beta_j|0, s_1)$ and $p(\beta_j|\gamma_j = 0, s_0) = de(\beta_j|0, s_0)$. Conditional posterior expectation of S_j^{-1} is derived as

$$E(S_j^{-1}|\beta_j) = E(\frac{1}{(1-\gamma_j)s_0 + \gamma_j s_1}|\beta_j) = \frac{1-p_j^g}{s_0} + \frac{p_j^g}{s_1} \tag{2.42}$$

Similarly, the M-step of the parameters (β, θ) can be updated by maximizing the posterior expectation of the log joint posterior density with γ_j and S_j^{-1} replaced by their conditional posterior expectations. The regression coefficient β is prepared by maximizing the expression as

$$Q_1(\beta) = pl(\beta) - \sum_{j=1}^{J} \bar{S}_j^{-1}|\beta_j| \tag{2.43}$$

and $\bar{S}_j^{-1}$ is presented as the conditional posterior expectation of S_j^{-1}.The partial likelihood derivatives with LASSO regression is presented in Figure 2.1.

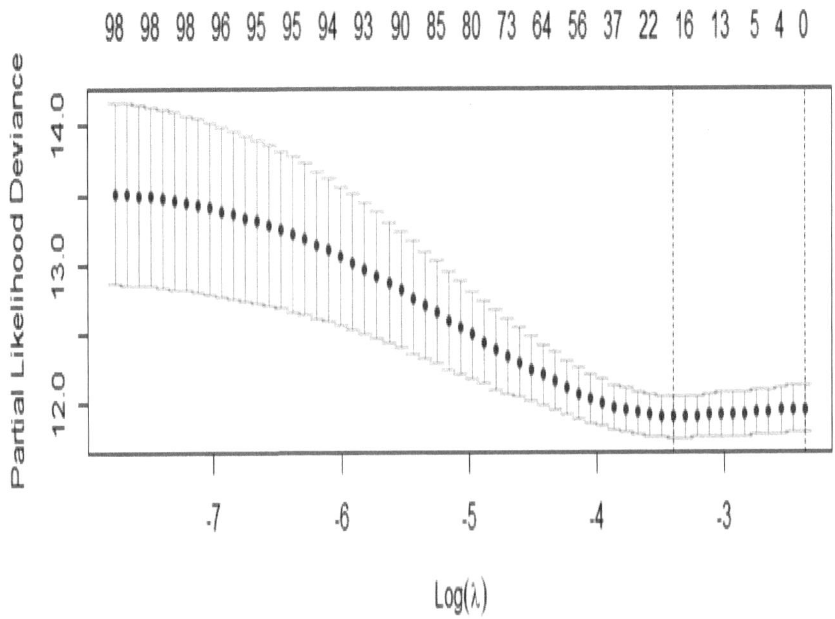

FIGURE 2.1: Lasso regression estimates on survival analysis.

Chapter 3

Parametric Survival Analysis

3.1 Introduction

The time-to-event regression analysis plays a crucial role in oncology research, and the application spread in different fields like biometrics, biostatistics, and bioinformatics. The application with high-dimensional data with time-to-event becomes challenging due to data complexity. Now genomic provides a large size of data with predictor variables. Sometimes it becomes more critical due to the availability of longitudinal data, adverse events, etc. All applications merge with high-dimensional data challenges. A large number of predictor variables and sample size both complex the high dimensional data analysis. But Cox model is widely applicable as predictor variables as well as large numbers of observations.

Simultaneously, the parametric survival model has become popular for analyzing survival data. It provides better predictive accuracy. This method makes to scale the data involvement for a large number of observations. Now computing the maximum likelihood, the parametric survival model needs a non-linear optimization problem. It helps with matrix inversion, solving large-scale problems by software.

There is a way to work with a regularized approach as a parametric choice of survival data analysis. It works by selecting the regularizing prior probability distribution with the model parameters to favor the fitted model and creating the point estimates as zero. However, traditional model fitting criteria get difficulties with overfitting problems to reduce the high dimensional variables. But the model calibration could be suitable to avoid the overfitting challenges. Survival analysis is not the new domain in statistics but still getting the opportunity for development with various applications and disciplines. Conventional

DOI: 10.1201/9781003185598-3

survival analysis is associated with the application of clinical trials and medical research. But this application spread to other disciplines like reliability engineering, demography, and sociology. However, the initial application never faced the challenges of working with a large number of variables.
The applications are focused on the small number of variable and their selection. Recent expansion with high dimensional makes the application of survival analysis real challenging. Other discipline areas of development with single nucleotide polymorphism (SNP) and gene expression microarray raised the requirement to expand this application. Now high dimensional work with survival analysis and parametric handling distribution is a challenging task. This chapter aims to present the application of survival analysis in high dimensional variable selection by parametric model.

3.2 Regularized Survival Analysis

Suppose the number of individuals are defined as n. Their corresponding survival time is defined as $y_i = \min(t_i, c_i), i = 1, ..., n$, where t_i and c_i . These are presented as failure time and right-censoring time for each individuals. Suppose $\delta_i = I(t_i \leq c_i)$ is presented as the indicator variable δ_i. Suppose $x_i = [x_{i1}, x_{i2}, ..., x_{ip}]^T$ is presented as p-vector of covariates. It is assumed that t_i and c_i are independent on x_i and respect to the censoring time as non-informative. The observed data can be presented as $D = \{(y_i, \delta_i, x_i) := 1, ..., n\}$. Let θ is presented as unknow through the model parameters. It can be presented as $y_1, y_2, ..., y_n$ as an identical and independent distribution by the survival function $f(y|\theta)$ and $S(y|\theta)$ respectively to parameterize by θ. Now the likelihood $L(\theta|D)$ of the parametric model is presented as

$$L(\theta|D) = \prod_{i=1}^{n} f(y_i|\theta)^{\delta_i} S(y_i|\theta)^{(1-\delta_i)} \tag{3.1}$$

The parametric models can present the distribution as Exponential, Weibull, Log-Logistic, or Log-Normal distributions. Each distribution can be formulated as a fully parametrized distribution pair by $\theta = (\lambda, \alpha)$. Now the parameter λ could be reparameterizedin terms of the covariates x and the vector β

such that

$$\lambda_i = \phi(\beta^T x_i), i = 1, ..., n \tag{3.2}$$

Finally, the likelihood function can be presented as

$$L(\beta, \alpha|D) = \prod_{i=1}^{n} f(y_i|\lambda_i, \alpha)^{\delta_i} S(y_i|\lambda_i, \alpha)^{(1-\delta_i} \tag{3.3}$$

The parameters β and α are obtained by maximizing the joint posterior density

$$L(\beta, \alpha|D) \propto L(\beta, \alpha|D)\pi(\beta)\pi(\alpha) \tag{3.4}$$

However, the joint posterior distribution is never generated with closed-form by (β, α). But it can be obtained as conditional posterior distribution $\pi(\beta|\alpha, D)$ and $\pi(\alpha|\beta, D)$ are concave and therefore can be solved efficiently. We can estimate β and α by maximizing the conditional posterior as

$$L(\beta, \alpha|D) \propto L(\beta, \alpha|D)\pi(\beta) \tag{3.5}$$

It is required to avoid the training data from the overfitting problem. Commonly, Bayesian could come and avoid overfitting by specifying the prior distribution on β as β_j becomes zero. This prior parameter selection is suitable for Gaussian and Laplacian prior categories. Bayesian could be used as a form of regularization in penalized manner.

3.3 Gaussian Prior and Ridge Regression

Now L_2 regularization works nicely. It assumed with the Gaussian prior for β_j having variance and mean with zero by τ_j as,

$$\pi(\beta_j|\tau_j) = N(0, \tau_j) = \frac{1}{\sqrt{2\pi\tau_j}} exp(-\frac{\beta_j^2}{2\tau_j}) \tag{3.6}$$

Now the mean zero with prior preference with value β_j is close to zero.

Now variance τ_j becomes positive through controlling the regularization with cross-validation. While τ_j becomes smaller, then it is assumed that β_j is being close to zero. If β_j becomes large, then it shows less informative prior. Simply, it can be assumed that $\tau_1 = \tau_2 = \tau_3 = ... = \tau_p$. Now β becomes as independent priori, the complete priori presented for each individual β_j, as $\pi(\beta|\tau_1, ..., \tau_p) = \prod_{j=1}^{p} \pi(\beta_j|\tau_j)$. Now the maximum posterior estimate β shows the prior equivalent with ridge regression. Perhaps, the Gaussian prior helps to consider the value of β_j as close to zero. But the posterior value of β_j varies from zero.

3.4 Laplacian Prior and Lasso Regression

Now we can assume that β_j follows the Gaussian distribution with mean zero and variance τ_j. Now in terms of L_1 regularization, it is assumed that β_j follows the Gaussian distribution having zero mean and variance τ_j. But τ_j can be fixed as exponential distribution parameterized by γ_j with the density

$$\pi(\tau_j|\gamma_j) = \frac{\gamma_j}{2}\exp(-\frac{\gamma_j}{2}\tau_j) \tag{3.7}$$

Now can be integrated by τ_j with equivalent non-hierarchical, double-exponential (Laplace) distribution with density as

$$\pi(\beta_j|\gamma_j) = \frac{\sqrt{\gamma_j}}{2}\exp(-\sqrt{\gamma_j|\beta_j|}) \tag{3.8}$$

It is easier to assume that $\gamma_1 = \gamma_2 = ... = \gamma_p$ to form the Gaussian prior for the components of β as independent.

3.5 Parameteric Survival Analysis

The widely used method is the non-parametric method, i.e., Kaplan-Meier or semi-parametric Cox model. Now the Cox model is defined as an extension of the baseline hazard model by considering the covariates. In contrast, the parametric survival analysis specified hazard function through consideration of different covariates. The estimation becomes more robust through the estimated survival curves as they become smoother to draw the information. It is possible to create a more sophisticated analysis by parametric modeling by including random effects through Bayesian inference. Perhaps, the parametric requires attention toward distributional assumption, and it is a tricky task sometimes. Parametric methods are easy to perform by partial-likelihood-based models. It helps to define the standard likelihood theory through parametric estimation and inference. Perhaps, accommodating complex censoring and truncation are easily linked with parametric models through partial likelihood methods. But the model validity depends heavily on parametric models. Different distributions like Weibull, gamma, and exponential could serve the parametric distributions, and these are commonly used parametric distributions.

3.6 Different Distribution

3.6.1 Exponential distribution

The Simplest form of the parametric distribution is the exponential distribution. It has a constant hazard function $h(t) = \lambda$, which gives the memoryless property. Now the risk of having the event of interest is similar to the point of the beginning. Now the pdf and survival functions are $f(t; \lambda) = \lambda e^{-\lambda t}$ and $S(t; \lambda) = e^{-\lambda t}$, respectively. The likelihood function is created by pdf of the survival function, and the right censoring mechanism can be included in it. The exponential distribution is an attractive choice with sample size. Now the survival modeling can be extended with Weibull distribution as a special case. If T follows the exponential distribution with parameter λ, then T^{α} becomes Weibull distribution with an additional parameter.

3.6.2 Weibull model

Now the Weibull survival function is expressed by hazard function as $h(t) = \alpha\lambda^{\alpha}t^{\alpha-1}$ and $S(t) = e^{-(\lambda t)\alpha}$. Now the Weibull distribution utilized with covariates on survival by the mean parameter $\mu = -\log\lambda$ and scale parameter $\sigma = 1/\alpha$ as

$$h(t) = \frac{1}{\sigma}e^{-\frac{\mu}{\sigma}}t^{\frac{1}{\sigma}-1} \tag{3.9}$$

Now this scale parameter is useful to make different formation of Weibull distribution. Now $\sigma = 1$ defines the Weibull distribution into exponential distribution with rate parameter λ. The complementary log-log $g(u) = log[-log(u)]$ transform it into the Weibull survival function as

$$\log[-\log(S_i)] = \alpha\log(\lambda) + \alpha\log(t_i) = -\frac{\mu}{\sigma} + \frac{1}{\sigma}\log(t_i) \tag{3.10}$$

by $S_i = S(t_i)$

$$S(t) = e^{-\frac{\mu}{\sigma}t^{\frac{1}{\sigma}}} \tag{3.11}$$

This outcome is useful as a diagnostic tool to understand how a set of survival data follows the Weibull distribution. Initially, the Kaplan-Meier estimator is estimated as e $y_i = \log\{-\log[\hat{S}(t_i)]\}$ and plot y_i versus $\log(t_i)$ toward utilizing the Weibull distribution.

3.6.3 Gamma distribution

Gamma distribution is a generalized form of the simple exponential distribution. The density function is presented as

$$f(t) = \frac{\lambda^{\alpha} t^{\alpha-1} e^{-\lambda t}}{\tau(\alpha)} \tag{3.12}$$

$$f(t) \propto t^{\alpha-1} e^{-\lambda t}$$

where

$$\tau(\alpha) = \int_0 |^{\infty} t^{\alpha-1} e^{-t} dt \tag{3.13}$$

is the Gamma function. For integer α, $\tau(\alpha) = (\alpha - 1)!$. There is no close formulae for survival or hazard function. $E(T) = \alpha/\lambda$. $\text{Var}(T) = \alpha/\lambda^2$. If $T_i \sim G(\alpha_i, \lambda), i = 1, \ldots\ldots, K$ and $T_i, i = 1, 2, , , , , , , K$ are independent , then

$$\sum_{i=1}^{K} T_i \sim G(\sum_{i=1}^{K} \alpha_i, \lambda) \tag{3.14}$$

$G(1, \lambda) \sim EXP(\lambda)$. $2\lambda G(K/2, \alpha) \sim \chi^2_K, \text{for} K = 1, 2, \ldots\ldots$

3.7 Maximum Likelihood Estimation

The Weibull distribution maximum likelihood can be estimated as

$$l(\lambda, \alpha) = \sum_{i=1}^{n} \{\delta_i \log[h(t_i)] + \log[S(t_i)]\} \tag{3.15}$$

The expression $h(t_i)$ and $S(t_i)$ becomes to

$$l(\lambda, \alpha) = d\log\alpha + d\alpha\log\lambda + (\alpha - 1)\sum_{i=1}^{n} \delta_i \log t_i - \lambda^{\alpha} \sum_{i=1}^{\alpha} t_i^{\alpha} \tag{3.16}$$

While $d = \sum_{i=1}^{n} \delta_i$. Further, the Weibull distribution assesses the effect of the covariates on survival by the usage of scale parameter $\sigma = \frac{1}{\alpha}$, and the mean parameter $\mu = -\log\lambda$. It is possible to calculate the maximum likelihood estimates by the parameters. The term can be created by $\alpha = 1/\sigma$ and $\lambda = e^{-\mu}$, and then compute the log-likelihood. Now the MLE could be obtained by the "optim" function through the estimates of μ and σ from the linear regression.

3.8 Illustration Using R

Estimates of the survival analysis with Exponential distribution presented in Table 3.1. Similarly, with Weibull distribution presented in Table 3.2.

Define Data

```
library(SurviMChd)
data(mcsurv)
survMC(4,8,2,10,mcsurv)
```

Listing 3.1: Exponential distribution fitting in survival analysis using R

```
library(survival)
data(cancer)
fit <- survreg(Surv(time, status) ~age + sex +
ph.karno, data=cancer,dist='exponential')
> summary(fit)

Call:
survreg(formula = Surv(time, status) ~ age + sex +
ph.karno,data = cancer, dist = ``exponential")
                 Value Std. Error     z        p
(Intercept)    5.07237    0.87371  5.81  6.4e-09
age           -0.01147    0.00928 -1.24   0.2168
sex            0.47085    0.16728  2.81   0.0049
ph.karno       0.01277    0.00596  2.14   0.0321

Scale fixed at 1

Exponential distribution
Loglik(model)= -1147.7   Loglik(intercept only)= -1156
        Chisq= 16.68 on 3 degrees of freedom, p= 0.00082
Number of Newton-Raphson Iterations: 4
n=227 (1 observation deleted due to missingness)
```

Install package for high dimensional survival data

```
library(SurviMChd)
data(headnneck)
```

Function with Exponential distribution

```
survexpMC(m1=8,n1=12,m2=4,n2=7,chains=2,iter=10,
data=headnneck)
```

Function with Weibull distribution

```
data(headnneck)
survweibMC(m1=8,n1=12,m2=4,n2=7,chains=20,iter=100,
data=headnneck)
```

TABLE 3.1: Survival analysis with Exponential distribution.

Parameter	Mu(SD)	2.5%	25%	50%	75%	97.5%	Rhat	n.eff
β_{11}	7.17(0.47)	6.58	6.86	7.15	7.34	7.97	2.31	3
β_{12}	−0.20(0.06)	−0.31	−0.25	−0.20	−0.16	−0.13	2.88	3
β_{13}	0.01(0.05)	−0.07	−0.00	0.02	0.05	0.08	1.56	5
β_{14}	0.39(0.45)	−0.14	−0.02	0.32	0.77	1.06	2.88	3
β_{15}	−0.25(0.28)	−0.65	−0.48	−0.21	0.01	0.05	2.45	3
β_{16}	−0.20(s0.23)	−0.47	−0.42	−0.23	0.03	0.09	1.97	4

TABLE 3.2: Survival analysis with Weibull distribution.

Parameter	Mu(SD)	2.5%	25%	50%	75%	97.5%	Rhat	n.eff
β_{11}	7.41(0.47)	6.65	7.22	7.48	7.61	8.09	0.90	10
β_{12}	−0.05(0.03)	−0.10	−0.07	−0.05	−0.03	−0.02	1.00	10
β_{13}	−0.01(0.03)	−0.06	−0.03	−0.01	0.00	0.02	1.39	6
β_{14}	0.07(0.34)	−0.43	−0.14	−0.00	0.36	0.57	1.13	10
β_{15}	0.01(0.20)	−0.37	−0.01	0.01	0.12	0.26	1.62	4
β_{16}	−0.07(0.15)	−0.31	−0.14	−0.09	0.01	0.15	1.25	10

Chapter 4

Competing Risk Modeling in High Dimensional Data

4.1 Introduction

Oncological event prediction is always a challenging task in improving the current treatment protocol. Unless event prediction becomes robust, it isn't very sensible to prepare protocol and treatment guidelines for future direction. It is to be noted that the risk of dying is not the same for all individuals. Neither is the response to treatment the same for all. A variety of factors, especially a patient's individuality arising out of their inherent genetic makeup, are to be held accountable for this differential. One option is to account for this variability by introducing random effects in our model, i.e., by using frailty models.

However, such modeling assumptions are more relevant in multivariate survival data. Another option is the identification of the potential biomarkers that are driving the entire process of cancer development. This information helps to decide what covariates should be included in the study.

There are different types of genomic data used for this purpose, like RNASeq, DNA Methylation, copy number variation, and microarray data. These high-dimensional data are also known as omics data. However, it is not only about building a statistical model from the omics data. Genomic data can only support individual-level risk score prediction or personalized medicine.

Similarly, survival data is always outcome data in the oncology domain. Thus, in this context, survival and genomic data come together to support personalized medicine and generate the individual-level risk score. The results of such a study are beneficial as reference material for prescribing an effective

DOI: 10.1201/9781003185598-4

line of treatment to patients who get diagnosed with a similar set of conditions and also for the purpose of drug development.

The joint work of survival data and high-dimensional data is not new. However, it is to be noted that high-dimensionality and survival analysis have their own set of challenges. Many of the problems are solved, and many of them are not. Mostly oncology domain is presented with follow-up observations like recurrence, response, and death. The time-to-occurrence of the event plays a crucial role in this direction. During the follow-up study, a patient's death may occur due to disease or other causes.

The reason is that although a lot of innovation has been achieved in the treatment procedures, which include very targetted and minimally invasive techniques for both diagnosis and treatment, patients are often affected by some secondary cardio-vascular or pulmonary disorders. Additionally, patients detected with carcinoma maybe already have some other health conditions.

This gives rise to a competing risk scenario. A competing risk is an event whose occurrence precludes the occurrence of the primary event of interest. Since this particular disease is associated with a very high rate of mortality, it is essential to categorize the total number of deaths into "deaths due to progression of cancer" and "deaths due to noncancerous causes." Naturally, the quality of treatment can be significantly improved if one takes into account the absolute risks of failure from different competing causes (different cause-specific risks). There are several applications with competing risk models in oncology research. However, there is a gap in the development of statistical methodology with competing risk survival models compatible with high-dimensional data, especially in the Bayesian scenario. In such a complicated scenario involving so many elements, Bayesian solutions based on prior information derived from expert opinion can often lead to very exact solutions. It is challenging to promote personalized medicine without having a robust statistical methodology to support it.

The objective of this chapter is to show the Bayesian risk prediction model in the survival competing risks for high dimensional Big cancer data. Conventionally, model prediction is performed by considering the overall cause of mortality. However, it must be remembered that mortality may be divided into cancer-specific deaths and deaths due to other competing causes.

Therefore, a robust predicting model can only be implemented by considering the interplay of risks, both due to the main as well as the competing cause of death. Moreover, it has been observed that the risk of mortality from other non-cancerous causes becomes heightened during the follow-up period.

The competing risk model is relatively standard in high-risk patients for a longer duration of follow-up. Their effect may be higher in older and palliative patients. However, the conventional statistical model does not consider the competing risk and performs survival data analysis by defining the cause of death only because of cancer.

The estimate of the predicted model becomes biased in the absence of competing risk. Hence the consideration of an excellent competing risk scenario in the analysis becomes necessary for correct inference.

There is abundant literature about competing risk concepts. Here we start by presenting the conventional statistical modeling associated with competing risk and extend the context of risk prediction in the case of survival analysis with high dimensional data.

While studying the survival of a group of subjects, it is often seen that certain underlying factors, such as age, sex, genomic profile, etc., ultimately affect the outcome of the study (i.e., the survival of these individuals). Only after proper adjustment corresponding to such explanatory variables does the survival analysis of a group of individuals become more accurate.

In particular, considers the genomic profiles of these individuals and holds them as potential factors affecting their survival. Such variables may be used as covariates (also known as explanatory variables, risk factors, and independent variables) in explaining the response or the dependent variable. The Cox proportional hazard model is an advantageous way in which covariates can be incorporated into any survival analysis.

It is possible to explore the application of high dimensional gene expression data in the context of competing for risk survival analysis. This results in covariate space becoming huge. Competing risk modeling coupled with gene expression profile makes this study interesting as well as challenging. Although analysis of gene expression data is not new, its implication in a competing risk survival scenario is still the lesser-explored domain.

It is to be noted that by considering omics data as covariates in this analysis, we are faced with a very typical situation which is popularly known as the "small p large n situation." As such, the conventional statistical methodology often fails to provide very satisfactory solutions in this situation. A unique classical estimate may not exist in such a situation.

The problem of dimension reduction can often become imperative for ease of analysis as well as to provide inferences that might ultimately be clinically useful. Many authors have studied the problem of dimension reduction. The most common solution while dealing with genomics data is to go for linear modeling along with ridge regression.

Additionally, one can propose a penalized estimator of the LASSO (Least absolute shrinkage and selection operator) due to this scenario. However, it is not stable for a large number of variables in microarray data . Owing to these problems, one can always look for Bayesian solutions that have options for dealing with high dimensional data, be it in dimensionality reduction or variable selection.

4.2 Survival and Competing Risk Model

In the absence of competing risk, the event time can be represented by a random variable T whose realized value is t. The probability density function (pdf) of T can be represented by $f(t)$. The probability of survival of the individual up to time t is given by the survival function

$$S(t) = P(T \geq t) = \int_t^{\infty} f(t)dt \tag{4.1}$$

The distribution function $F(t)$ represented by $F(t) = 1 - S(t) = P(T \leq t)$ shows the incidence of the event of the duration of the follow-up. Similarly, the hazard function is presented as

$$h(t) = \lim_{\Delta t \to 0} \frac{P(t \leq T < t + \Delta T | T \geq t)}{\Delta t} \tag{4.2}$$

The hazard as a function of time defines the instantaneous rate of appearance of the event of interest in individuals who are at risk of the event. The cumulative hazard function is presented as

$$H(t) = \int_0^t h(u)du \tag{4.3}$$

Further, the survival and hazard functions are linked by the cumulative hazard function as

$$S(t) = \exp(-H(t)). \tag{4.4}$$

The effect of concomitant variables can be introduced in the survival analysis by use of different relationships, among which the Cox proportional hazards model is the most famous. The Cox proportional hazards regression model, as is also obvious from the name, is based on the assumption of proportional hazard. It is specified as

$$\log(h(t)) = \log(h_0(t)) + \beta' Z \tag{4.5}$$

The term $h_0(t)$ defines baseline hazard function. The explanatory variables are presented as Z. Corresponding regression parameters are presented as $\beta = [\beta_1, \ldots, \beta_p]'$. It directly relates to the incidence of events and regression coefficient in the absence of competing risk. Now the survival function under the proportional hazards set up is represented as

$$S(t) = \exp\{-H(t)e^{\beta' Z}\} = S_0(t)^{\exp(\beta' Z)} \tag{4.6}$$

If the competing risk is not present, then the equation makes an inference about the covariate on the hazard function that allows for similar inference toward the effect of the covariates of the predictive model. In the analysis of lifetime data, various parametric models are used. Among these, a few are most widely used due to their flexibility and usefulness in explaining a wide variety of life situations. The Exponential and Weibull are two such models.

The exponential distribution has been one of the first and most widely used lifetime models. Although the assumption of constant hazard may be a little restrictive in certain lifetime situations, the easy interpretability and the existence of closed-form of statistical solutions for this model more than makes up for it. The p.d.f. and survival function of this model is given by $f(t) = \eta \exp(-\eta t)$ and $S(t) = \exp(-\eta t)$. The hazard function of the exponential is η. Considering the Cox-PH relationship, the p.d.f. and hazard functions for the exponential model are defined as

$$f(t|\eta, \kappa, \beta) = \eta \exp(\beta' Z) \exp\left(-\eta \exp(\beta' Z) t\right) \tag{4.7}$$

$$h(t|\eta, \beta) = \eta \exp\left(\beta' Z\right) \qquad \eta > 0, t > 0. \tag{4.8}$$

The Weibull distribution is perhaps one of the most flexible lifetime models in terms of accommodating a wide variety of hazard behaviors. Thus, it is used to represent the lifetime of numerous physical and biological entities. The p.d.f. and survival function of this model is given by $f(t) = \kappa\eta(\eta t)^{\kappa-1}\exp\left[-(\eta t)^{\kappa}\right]$ and $S(t) = \exp\left[-(\eta t)^{\kappa}\right]$. Considering the Cox-PH relationship, the p.d.f. and hazard functions for the Weibull model are presented as

$$h(t|\eta, \kappa, \beta) = \left\{\kappa\eta t^{\kappa-1}\right\}\exp\left(\beta' Z\right) \qquad \eta > 0, \kappa > 0, t > 0; \tag{4.9}$$

$$f(t|\eta, \kappa) = \kappa\eta t^{\kappa-1}\exp(\beta' Z)\exp\left(-\exp(\beta' Z)\eta t^{\kappa}\right) \tag{4.10}$$

It is to be noted that the primary outcome of this study is death due to cancer. Now those individuals who continue living even after the termination of the experiment or those that are lost to follow-up are known to have their lifetimes censored. Such an individual whose life is censored at a specific time point is still at risk of dying at a particular time point in the future. But we are not able to observe this death within our observation window.

4.3 The Competing Risk Models

Suppose we are studying the lifetime of an individual i, such that there are competing causes that might lead to his/her death. That is, we will end up observing either the survival time t_i or the censoring time c_i. Hence the observed time x_i of the individual can be denoted by the general representation $x_i = \min(t_i, c_i)$. It is to be noted that δ_i is a common indicator that represents censoring for $\delta_i = 0$ and $\delta_i = j$ indicates failure due to the jth cause where $j = 1, ..., J$. Since there are more than one causes of failure, one must take into account the cause specific hazard given by

$$h_j(t) = \lim_{\Delta t \to 0}\frac{\text{Prob}(t \leq T < t + \Delta T, \delta = j|T \geq t)}{\Delta t}, \tag{4.11}$$

where $h_j(t)$ is the cause specific hazard due to the jth cause. Thus the overall survival function becomes

$$S(t) = \exp\left[-\sum_{j=1}^{J}\int_0^t h_j(u)du\right]. \tag{4.12}$$

The cumulative incidence function for cause j are given by $F_j(t) = P(T \leq t, \delta = j)$. When we apply the context of the Cox-PH model for $h_j(t)$, we get

$$h_j(t|x) = h_{j_0}(t)\exp(\beta_j' Z). \tag{4.13}$$

Inference for the cause-specific hazards model with the proportional hazards assumption can be made by applying the Cox model to each cause. We have assumed that if patients die, followed by the relapse of the disease, death can be defined as death due to therapeutic failure and coded as status = 1. If patients followed as relapse-free but died, death is defined as death due to other causes and coded as status = 2. Patients who are alive can be considered as censored observations, and they can be coded as status = 0. Assuming that there are two causes of death, the competing risk model can be expressed as

$$\begin{aligned} L(t_1,\ldots,t_n|\beta_1,\beta_2) = \\ \prod_{i=1}^{n}\left[h_{10}(t_i)\exp(\beta_1' Z_i)\exp\left(-\int_0^{t_i} h_{10}(u)\exp(\beta_1' Z_i)du\right)\right]^{I(\delta_i=1)} \\ \times\prod_{i=1}^{n}\left[h_{20}(t_i)\exp(\beta_2' Z_i)\exp\left(-\int_0^{t} h_{20}(u)\exp(\beta_2' Z_i)du\right)\right]^{I(\delta_i=2)} \\ \times\prod_{i=1}^{n}\left[\exp\left(-\int_0^{t_i} h_{10}(u)\exp(\beta_1' Z_i)du - \int_0^{t} h_{20}(u)\exp(\beta_2' Z_i)du\right)\right]^{I(\delta_i=0)} \end{aligned} \tag{4.14}$$

After a little bit of manipulation, the above expression can be simplified as

$$\begin{aligned} L(t_1,\ldots,t_n|\beta_1,\beta_2) = \prod_{i=1}^{n}\left[h_{10}(t)\exp(\beta_1' Z)\right]^{I(\delta_i=1)} \\ \times\exp\left(-\int_0^{t} h_{10}(t)\exp(\beta_1' Z)dt\right) \\ \times\prod_{i=1}^{n}\left[h_{20}(t)\exp(\beta_2' Z)\right]^{I(\delta_i=2)}\exp\left(-\int_0^{t} h_{20}(t)\exp(\beta_2' Z)dt\right) \end{aligned} \tag{4.15}$$

It is observed that the regression coefficients for the two causes are separated symmetrically , and that for estimating each covariate, failed observations from the other cause play the same role as censored observations. Thus, inferential procedures can be conducted by treating deaths from other causes as censored. In other words, inferences can be drawn by fitting four separate proportional hazards models. If we assume Weibull baseline hazard for the both causes i.e.,

$$h_{10}(t) = \kappa_1 \eta_1 t_i^{\kappa_1 - 1} \tag{4.16}$$

$$h_{20}(t) = \kappa_2 \eta_2 t_i^{\kappa_2 - 1} \tag{4.17}$$

The cause specific hazards will be

$$h_1(t|x) = \kappa_1 \eta_1 t_i^{\kappa_1 - 1} \exp(\beta_1' Z_i) \tag{4.18}$$

and

$$h_2(t|x) = \kappa_2 \eta_2 t_i^{\kappa_2 - 1} \exp(\beta_2' Z_i) \tag{4.19}$$

respectively. The corresponding likelihood function under competing risk can be expressed as

$$\begin{aligned} L_{WW}(t;\kappa_1,\kappa_2,\eta_1,\eta_2,\beta_1,\beta_2) = \prod_{i=1}^{n} \big[\kappa_1 \eta_1 t_i^{\kappa_1 - 1} \\ \exp(\beta_1' Z_i)\big]^{I(\delta_i = 1)} \exp\left(-\exp(\beta_1' Z_i)\eta_1 t_i^{\kappa_1}\right) \\ \times \prod_{i=1}^{n} \left[\kappa_2 \eta_2 t_i^{\kappa_2 - 1} \exp(\beta_2' Z_i)\right]^{I(\delta_i = 2)} \exp\left(-\exp(\beta_2' Z_i)\eta_2 t_i^{\kappa_2}\right) \end{aligned} \tag{4.20}$$

Similarly, if we assume exponential baseline hazard for both causes i.e., $h_{10}(t) = \eta_1$ and $h_{20}(t) = \eta_2$, the cause specific hazards will be $h_1(t|x) = \eta_1 \exp(\beta_1' Z_i)$ and $h_2(t|x) = \eta_2 \exp(\beta_2' Z_i)$, respectively. Thus, the likelihood under the assumption of competing risk can be expressed as

$$\begin{aligned} L_{EE}(t;\eta_1,\eta_2,\beta_1,\beta_2) = \prod_{i=1}^{n} \left[\eta_1 \exp(\beta_1' Z_i)\right]^{I(\delta_i = 1)} \exp\left(-\exp(\beta_1' Z_i)\eta_1 t_i\right) \\ \times \prod_{i=1}^{n} \left[\eta_2 \exp(\beta_2' Z_i)\right]^{I(\delta_i = 2)} \exp\left(-\exp(\beta_2' Z_i)\eta_2 t_i\right) \end{aligned} \tag{4.21}$$

If we assume exponential baseline hazard for the first cause i.e., $h_{10}(t) = \eta_1$, and weibull baseline hazard for the second cause i.e., $h_{20}(t) = \kappa_2 \eta_2 t_i^{\kappa_2 - 1}$, the cause specific hazard can be expressed as $h_1(t|x) = \eta_1 \exp(\beta_1' Z_i)$ and $h_2(t|x) = \kappa_2 \eta_2 t_i^{\kappa_2 - 1} \exp(\beta_2' Z_i)$, respectively. Thus, the likelihood under the assumption of competing risk can be given by

$$\begin{aligned} L_{EW}(t; \kappa_2, \eta_1, \eta_2, \beta_1, \beta_2) &= \prod_{i=1}^{n} \left[\eta_1 \exp(\beta_1' Z_i)\right]^{I(\delta_i = 1)} \exp\left(-\exp(\beta_1' Z_i)\eta_1 t_i\right) \\ &\times \prod_{i=1}^{n} \left[\kappa_2 \eta_2 t_i^{\kappa_2 - 1} \exp(\beta_2' Z_i)\right]^{I(\delta_i = 2)} \exp\left(-\exp(\beta_2' Z_i)\eta_2 t_i^{\kappa_2}\right) \end{aligned} \tag{4.22}$$

Similar to the above assumption, if we consider Weibull baseline hazard for the first cause i.e., $h_{10}(t) = \kappa_1 \eta_1 t_i^{\kappa_1 - 1}$, and exponential baseline hazard for the second cause i.e., $h_2(t|x) = \eta_2$, respectively. The likelihood under the assumption of competing risk can be given by

$$\begin{aligned} L_{WE}(t; \eta_1, \kappa_1, \eta_2 \beta_1, \beta_2) &= \prod_{i=1}^{n} \left[\kappa_1 \eta_1 t_i^{\kappa_1 - 1} \exp(\beta_1' Z_i)\right]^{I(\delta_i = 1)} \\ &\quad \times \exp\left(-\exp(\beta_1' Z_i)\eta_1 t_i^{\kappa_1}\right) \qquad (4.23) \\ &\times \prod_{i=1}^{n} \left[\eta_2 \exp(\beta_2' Z_i)\right]^{I(\delta_i = 2)} \exp\left(-\exp(\beta_2' Z_i)\eta_2 t_i\right) \end{aligned}$$

In order to draw inferences about the competing risk model, one of the most valuable procedures is a method of maximum likelihood. Since we are working on Gene Expression data, we have more than 20K covariates. Obviously, we have a number observation (n) less than the number of covariates (p), and thus, MLE is not feasible in this case, and no unique MLE can be obtained. Even Bayesian analysis with a high number of covariates will not provide any significant conclusion.

Thus, our next objective is to reduce the number of covariates without losing too much information. There are a variety of articles available in the literature that suggests a number of methods to reduce the dimension of the covariate. In this article, we propose to use one of the popular methods, the least absolute shrinkage, and selection operator, to reduce the number of covariates. The following section will provide a brief discussion of the proposed Bayesian method and its implementations.

4.4 Aalen's Additive Hazards Model

The additive hazards model is defined as

$$\alpha_{0j;i}(t; Z) = \alpha_{0j;0}(t) + \beta_{0j}(t).Z_i, j = 1, 2, ...J, i = 1, ..., n \tag{4.24}$$

The term $\beta_{0j}(t)$ is presented with $1 \times p$ regression coefficient function with Z_i as the $p \times 1$ vector of covariates for individual i, and $\alpha_{0j;0}(t)$ is an unspecified, non-negative baseline hazard function. Similarly, the cumulative cause-specific hazard can be presented as $A_{0j;0}(t) + Z_i B_{oj}(t)$ where

$$A_{0j;0}(t) = \int_0^t \alpha_{0j;0}(u)du \text{ and } B_{0j}(t) = \int_0^t \beta_{0j}(u)du \tag{4.25}$$

Aalen proposes model (1980) [4] and that is studied by several others. One of the limitations of the Aalen estimator is that it is closed under addition [5]. So the cause-specific hazards are additive in nature, and the total hazard of dying is presented as $\sum_{j=1}^{J} \alpha_{0j;i}(t; Z_i)$. Sometimes, the standard least square estimator of the cumulative hazard presents a nice property.

The covariates are the same for all causes. The estimator of the total hazard for mortality is equivalent to the sum of the estimators of the cause-specific hazards. With the standard least-square estimators of $A_{0j;0}(t)$ and $B_{0j;0}(t)$, we can define it as $\hat{A}_{0j;0}(t)$ and $\hat{B}_{0j;0}(t)$. So it predicts the cumulative cause-specific hazards for a covariate vector equal to z as

$$\hat{A}_{0j}(t; z) = \hat{A}_{0j;0}(t) + z\hat{B}_{0j}(t) \tag{4.26}$$

Sometimes, this model can be adopted toward survival probability prediction and cumulative incidence functions by replacing the increments of the cause-specific Nelson-Aalen estimator. The predictor survival probability comes to

$$\hat{P}(T > t|z) = \prod_{s \leq t}(1 - (\sum_{j=1}^{J} \Delta\hat{A}_{0j}(s; z)) \tag{4.27}$$

The corresponding cumulative incidence function presented with

$$\hat{P}(T \leq t, X_T = j|z) = \sum_{s \leq t} \hat{P}(T > s - |z)/\Delta\hat{A}_{0j}(t; z) \tag{4.28}$$

4.5 Bayesian Formulation

In order to perform Bayesian analysis, we engage in a two-fold procedure. The first is model selection using the concept of shrinkage, and the second is an evaluation of the model-specific parameters with respect to the selected covariates. For both purposes, different sets of priors are formulated, and the Metropolis algorithm is employed. For the purpose of model selection, multivariate normal prior is used with mean vector and variance-covariance matrix whose dimension equals the dimension of the total parameter space (no. of model parameters + no. of covariates). After the variable selection procedure is complete, for the second step, we have a different set of priors for model evaluation purposes. The shape parameter of the Weibull distribution lies on a buoyant line. Thus, we have considered gamma prior to the shape parameters of Weibull distribution. The prior distribution can be expressed as

$$\pi(\kappa_1|a_1, b_1) = \frac{b_1^{a_1}}{\Gamma(a_1)}\kappa_1^{a_1-1}\exp(-b_1\kappa_1) \tag{4.29}$$

$$\pi(\kappa_2|a_2, b_2) = \frac{b_2^{a_2}}{\Gamma(a_2)}\kappa_1^{a_1-1}\exp(-b_1\kappa_2) \tag{4.30}$$

$$\pi(\eta_1|c_1, d_1) = \frac{d_1^{c_1}}{\Gamma(c_1)}\eta_1^{c_1-1}\exp(-d_1\eta_1) \tag{4.31}$$

$$\pi(\eta_2|c_2, d_2) = \frac{d_2^{c_2}}{\Gamma(c_2)}\eta_1^{c_1-1}\exp(-d_2\eta_2) \tag{4.32}$$

The priors for covariates are assumed to follow a p-variate normal distribution with mean zero vector and variance-covariance matrix $D_1(D_2)$, for first and second cause, respectively.

$$\pi(\beta_1|D_1) = \frac{1}{|2\pi D_1|^{\frac{p}{2}}}\exp\left(-\frac{\beta_1' D_1^{-1}\beta_1}{2}\right) \tag{4.33}$$

$$\pi(\beta_2|D_2) = \frac{1}{|2\pi D_2|^{\frac{p}{2}}}\exp\left(-\frac{\beta_2' D_2^{-1}\beta_2}{2}\right) \tag{4.34}$$

If we assume Weibull baseline hazard for the both causes i.e.,

$$h_{10}(t) = \kappa_1 \eta_1 t_i^{\kappa_1 - 1} \tag{4.35}$$

and

$$h_{20}(t) = \kappa_2 \eta_2 t_i^{\kappa_2 - 1} \tag{4.36}$$

the corresponding posterior is given by

$$\begin{aligned} p_{WW}(t; \kappa_1, \kappa_2, \eta_1, \eta_2, \beta_1, \beta_2) = \prod_{i=1}^{n} \left[\kappa_1 \eta_1 t_i^{\kappa_1 - 1} \exp(\beta_1' Z_i)\right]^{I(\delta_i = 1)} \\ \exp\left(-\exp(\beta_1' Z_i)\eta_1 t_i^{\kappa_1}\right) \\ \times \prod_{i=1}^{n} \left[\kappa_2 \eta_2 t_i^{\kappa_2 - 1} \exp(\beta_2' Z_i)\right]^{I(\delta_i = 2)} \exp\left(-\exp(\beta_2' Z_i)\eta_2 t_i^{\kappa_2}\right) \\ \times \pi(\kappa_1 | a_1, b_1) \times \pi(\kappa_2 | a_2, b_2) \times \pi(\eta_1 | c_1, d_1) \times \pi(\eta_2 | c_2, d_2) \\ \times \pi(\beta_1 | D_1) \times \pi(\beta_2 | D_2). \end{aligned} \tag{4.37}$$

If we assume exponential baseline hazard for both causes i.e., $h_{10}(t) = \eta_1$ and $h_{20}(t) = \eta_2$, the posterior is given by

$$\begin{aligned} p_{EE}(t; \eta_1, \eta_2, \beta_1, \beta_2) = \prod_{i=1}^{n} \left[\eta_1 \exp(\beta_1' Z_i)\right]^{I(\delta_i = 1)} \\ \exp\left(-\exp(\beta_1' Z_i)\eta_1 t_i\right) \\ \times \prod_{i=1}^{n} \left[\eta_2 \exp(\beta_2' Z_i)\right]^{I(\delta_i = 2)} \\ \exp\left(-\exp(\beta_2' Z_i)\eta_2 t_i\right) \\ \times \pi(\kappa_1 | a_1, b_1) \times \pi(\kappa_2 | a_2, b_2) \times \pi(\eta_1 | c_1, d_1) \\ \times \pi(\eta_2 | c_2, d_2) \times \pi(\beta_1 | D_1) \times \pi(\beta_2 | D_2). \end{aligned} \tag{4.38}$$

If we assume exponential baseline hazard for for the first cause i.e., $h_{10}(t) = \eta_1$, and Weibull baseline hazard for the second cause i.e., $h_{20}(t) = \kappa_2 \eta_2 t_i^{\kappa_2 - 1}$, the posterior under the assumption of competing risk can be given by

$$\begin{aligned} p_{EW}(t;\kappa_2,\eta_1,\eta_2,\beta_1,\beta_2) &= \prod_{i=1}^{n} [\eta_1 \exp(\beta_1' Z_i)]^{I(\delta_i=1)} \exp\left(-\exp(\beta_1' Z_i)\eta_1 t_i\right) \\ &\times \prod_{i=1}^{n} \left[\kappa_2 \eta_2 t_i^{\kappa_2-1} \exp(\beta_2' Z_i)\right]^{I(\delta_i=2)} \exp\left(-\exp(\beta_2' Z_i)\eta_2 t_i^{\kappa_2}\right) \\ &\times pi(\kappa_1|a_1,b_1) \times |pi(\kappa_2|a_2,b_2) \times \pi(\eta_1|c_1,d_1) \times \pi(\eta_2|c_2,d_2) \\ &\times \pi(\beta_1|D_1) \times \pi(\beta_2|D_2). \end{aligned} \tag{4.39}$$

Similar to the above assumption, if we consider Weibull baseline hazard for the first cause, i.e., $h_{10}(t) = \kappa_1\eta_1 t_i^{\kappa_1-1}$, and exponential baseline hazard for the second cause, i.e., $h_2(t|x) = \eta_2$, respectively, the posterior under the assumption of competing risk can be given by

$$\begin{aligned} p_{WE}(t;\eta_1,\kappa_1,\eta_2\beta_1,\beta_2) &= \prod_{i=1}^{n} \left[\kappa_1 \eta_1 t_i^{\kappa_1-1} \exp(\beta_1' Z_i)\right]^{I(\delta_i=1)} \\ &\times \exp\left(-\exp(\beta_1' Z_i)\eta_1 t_i^{\kappa_1}\right) \\ \times \prod_{i=1}^{n} [\eta_2 \exp(\beta_2' Z_i)]^{I(\delta_i=2)} &\exp\left(-\exp(\beta_2' Z_i)\eta_2 t_i\right) \\ &\times pi(\kappa_1|a_1,b_1) \times |pi(\kappa_2|a_2,b_2) \times \pi(\eta_1|c_1,d_1) \\ &\times \pi(\eta_2|c_2,d_2) \times \pi(\beta_1|D_1) \times \pi(\beta_2|D_2). \end{aligned} \tag{4.40}$$

4.6 The Lasso Method

The least absolute shrinkage and selection operator(Lasso) is a shrinkage and variable selection method primarily used in linear regression. The process introduces a specified upper bound to the sum of the absolute value of parameters. This causes the regression coefficients of certain regressors to shrink toward zero. This very much eases the process of variable selection as it helps to identify the most significant predictors for the response variable. Thus the process compacts the model to a simpler and more meaningful form, even reducing prediction errors to a great extent.

In other words, LASSO removes vain features from the models. In the case of positively correlated high dimensional data, it generates the correlated features. It is commonly used in feature selection technique and uses the $L_1 = \sum_j |\beta_j|$ penalty term instead of $L_2 = \sum_j \beta_j^2$, which is traditionally used in another similar procedure called ridge regression. Here β_j's are regressor coefficients. It penalized the factor toward least square or as maximum likelihood estimation.

It consists of tuning constant λ_L, which is multiplied with the penalty term, i.e., the sum of absolute numbers as the parameter of interest β. The tuning constant is the penalizing factor that helps to decide how many predictors would make it to the final list. The formula becomes $\lambda_L \sum_{j=1}^{p} |\beta_j|$. If $(j = 0, 1, \ldots, p)$ then the LASSO penalized function say L_{LASSO} becomes

$$L_{LASSO} = \log L - \lambda_L \sum_{j=1}^{p} |\beta_j| \tag{4.41}$$

Final results and inferences are quite sensitive to the variation in λ. Thus it must be carefully determined. We preferred using the LASSO as the initial feature selection technique irrespective of correlated/non-correlated features.

The intention to use LASSO was to reduce the burden of high dimensional data and proceed with the application of parametric Bayesian. There are other variable selection techniques as well. However, we skip this discussion as the focus of the paper is not to compare various variable selection tools.

4.7 Metropolis Algorithm

The Metropolis is a robust Markov chain-based algorithm used mainly to simulate multivariate distributions that are often complex or non-standard. The algorithm proceeds by creating a Markov chain such that simulating from its invariant density would finally render samples from the target posterior.

Thus implementation of the Metropolis algorithm requires generating from the proposal density and then by using a randomization step, either accepting or rejecting the generated variate values. A variate value is acceptable if it results in a sizeable posterior probability. However, if the generated value is left, the algorithm retains the value generated at the previous step.

Successively running the Metropolis algorithm provides an iterating chain of variates that should finally converge to a random sample from the concerned posterior. Once the convergence is achieved, one can pick up the generated observations at certain equidistant points to form a random sample of appropriate size from the desired posterior to cut down on the autocorrelation.

A critical question is how to choose the center and the spread of the normal kernel so that it is clearly optimal for the generation of variate values. The efficiency of the algorithm, of course, depends on this choice.

If the spread is too large, then a lot of such points may be generated that are rather far away from the concerned posterior. On the other hand, the spread is too much narrowed down; the chain might take a longer time to traverse the entire parameter space with low probability points being undersampled.

A better suggestion is to approximate the mean and variance of the corresponding kernel by the maximum likelihood estimates and the corresponding Hessian-based approximation. In actual practice, we also make use of a tuning constant c_0 for adjusting the Hessian.

The value of c_0 often varies from 0.5 to 0.9. Another option would be to use sample-based approximations for the scaling component of the kernel, viz., the Hessian, after the first few iterations. This shall increase the efficiency of the generating algorithm.

4.8 Deviance Information Criterion and Akaike Information Criteria

This is a Bayesian model comparison criterion that includes terms that reward a model for better fit and another that penalizes the same model for an increase in complexity.

The criterion strives to achieve a very judicious trade-off between the two extremes. Thus it uses the posterior mean of deviance and adds a penalty term for complexity. It is given by

$$DIC = \overline{D(\theta)} + 2p_{D_j} \tag{4.42}$$

where $D(\theta) = -2\log(L(t|\theta)) + 2\log(L(t|\hat{\theta}))$ is the deviance, $\hat{\theta}$ is the posterior estimate of the parameter (or parameter vector), and $p_{D_j} = \overline{D(\theta)} - D(\bar{\theta})$ is the penalty term. For details on the application of the Deviance Information Criterion (DIC) to statistical models one can refer to among others. The Akaike Information Criteria is another r commonly used model selection tool that is given by

$$AIC = 2k - 2\ln L \tag{4.43}$$

where k is the number of parameters in the model and L is the likelihood.

4.9 Illustration with Example Data

Handling high-dimensional data is obviously one of the most challenging practical problems in the analysis of any microarray data. It comes with several thousands of variables, and running so many variables for regression analysis is a difficult task. Here we presented the "SurviMChd" package that can handle several thousands of variables simultaneously. We prepared a simple dataset having normal distribution to simply explain the core area of the function and its application. Perhaps, it can be extended to any number of variables, provided the dataset has a "Death" column to define the death status of the patients as 1(if dead) or o(if alive). In addition, the overlall survival(OS) variable is required to be presented as positive values by the duration of survival of the patients.

Further, to extend the application with competing risk, the dataset requires to have a competing risk column as the variable. Now patients die due to other causes; then, it will be presented as 2. Now working the entire column with competing risks is not suitable in the context of survival analysis. So entire example data set is split into two parts, exdata2a, and exdata2b.

Now results obtained on both the risk of death are presented below. It is to be noted that the total number of possible submodels that can be formulated using a set of regressors is 2^p, where p is the number of regressors. An essential problem in statistical analysis is the choice of an optimal model from a set of plausible models, which ultimately boils down to the selection of the appropriate subset of variables to be included.

It is a vital step, especially in the analysis of omics data, as it is typically seen that a relatively small number of explanatory variables (genes in this case) explains a significantly large part of the variation in the response variable, in this case, the lifetime of patients. If we were to employ some variable selection method that is based on the computation of some model selection criteria like DIC, we would have exponentially increased our computational requirements with no significant gain in information.

Such methods are feasible to be used only in situations where the choice is, say, among a dozen models. In a Bayesian framework, the model selection process transcends to calculating the posterior probability of each model, which, in keeping with the concepts discussed earlier, transforms to estimating the marginal posterior probability that a particular regressor should be included in the model.

But, that again would not be feasible in case the number of predictors is vast. So we use the concept of shrinkage, first through LASSO and then through the Bayesian equivalent of ridge regression by running a Metropolis sampler using a normal prior with mean zero and variance-covariance matrix equal to identity matrix of dimension equal to the number of predictors in the study.

Thus we have conducted a two-step variable selection process, first through LASSO and then through the Metropolis-Hastings algorithm. To clarify, we have run a LASSO algorithm primarily for the set of all 22283 regressors and found around 400 to be significantly crucial for our analysis. One can use any of the available algorithms or R packages to obtain the same.

The R package SurvHiDim can be an excellent option to run LASSO. Next, we have run the Metropolis algorithm in accordance with the description provided to obtain the estimates for all of these 400 parameters with multivariate normal prior.

Each element of the mean vector is zero, and the variance-covariance matrix is the identity matrix. For the purpose of generating samples from the resultant posterior surface, we use the Metropolis sampler. For this purpose, we consider a multivariate normal kernel density. The mean vector is estimated by the maximum likelihood estimators and the variance-covariance matrix by a Hessian-based approximation. These are obtained using the "survreg" function of the "survival" package in R. The parameter estimates obtained from running the algorithm provides us with the decision whether or not to include those variables in the final regressor space.

We have removed those regressors whose coefficient estimates shrunk nearly to zero from the set of regressors for the final analysis. After a careful examination of the results obtained from running the Metropolis algorithm for all the four models, namely, exponential-exponential (EE), exponential-Weibull (EW), Weibull-exponential (WE), and Weibull-Weibull (WW), we have concluded with a set of 14 genes (regressors) that seem to have a significant effect on the clinical condition of the patients. Applied practitioners may, however, skip this computationally intensive technique and go for feature selection in high dimensional data using one or other R package of which the highMLR is a recent and practical example.

However, different sensitive variable selection techniques can be adopted, and we can compare their performance by model diagnostics criteria. But the primary goal of this work was not about the variable selection. In order to analyze our data in competing risk scenarios, we have used the clinical characteristics to define the two risks. We have assumed that if patients die, followed by the relapse of the disease, death can be defined as death due to therapeutic failure and coded as status = 1. If patients followed as relapse-free but died, death is defined as death due to other causes and coded as status=2. Patients who are alive can be considered as censored observations, and they can be coded as status = 0. The competing cause can be separated in an above-mentioned manner.

It is to be noted that a significant aim of this research is variable selection in the presence of a competing risk scenario. Or in other words, it is mainly our aim to determine which are the driving genes that are affecting the probability of survival of a patient having adenocarcinoma at present or as clinical history.

Additionally, it is to be kept in mind that a patient who dies may die due to carcinoma or due to allied causes creating a competing risk scenario. From all the four considered models, it is seen that the gene ApoE is the most potent, followed by PDE12, RAD23B, APC, SYNE1, and DOCK2. Apart from these genes, age is a vital factor affecting the patient's survival probability.

In the second step, Metropolis algorithm has been applied to the study the posteriors 4.37, 4.38, 4.39, and 4.40. It is to be noted that the computation for the Metropolis algorithm can be carried out by considering only one cause at a time and by treating deaths due to other causes as censored observations.

We have used the function survreg from the package survival to calculate the maximum likelihood estimates and the Hessian matrix. Next, a multivariate normal kernel can be employed to run the Metropolis algorithm for each of the posteriors 4.37, 4.38, 4.39, and 4.40.

The mean vector of this kernel has been estimated by the maximum likelihood estimates and the variance-covariance matrix by the Hessian. After the stationarity of the Markov chain generated from the algorithm is observed at iterations, equispaced samples from the chain are deemed to be independent samples from the concerned posteriors.

Sample-based estimates of every posterior characteristic can now be calculated and are given in the tables. The posterior mean, median, and mode, along with the highest probability density intervals, have been provided in these tables. The classical estimate MLE is also provided for the purpose of reference.

Once the genes are chosen for the analysis by the Bayesian methodology, there is a clear expectation that these genes shall affect the outcome variable. A suitable strategy, in this case, may therefore be to fit the full model and then interpret the sizes of the posterior estimates of the parameters in terms of their importance.

4.10 Bayesian for Competing Risk Analysis Illustration Using R

Preparation of example data

```
library(SurviMChd)
e1=rnorm(100,5,0.5)
e2=rnorm(100,10,0.6)
e3=rnorm(100,15,2)
e4=rnorm(100,20,2)
f1=rnorm(100,8,2)
f2=rnorm(100,9,2)
f3=rnorm(100,11,2)
f4=rnorm(100,13,2)
f5=rnorm(100,10,2)
Death=rbinom(100,1,.5)
OS=rnorm(100,100,20)
cmprsk=rbinom(100,2,0.5)
exdata1=data.frame(e1,e2,e3,e4,f1,f2,f3,f4,f5,Death,OS,cmprsk)
```

Bayesian Exponential distribution fitting on survival data

```
survexpMC(m1=5,n1=9,m2=1,n2=4,chains=2,iter=10,data=exdata1)
```

Listing 4.1: Output on Bayesian Exponential distribution

	mu.vect	sd.vect	2.5%	97.5%	Rhat	n.eff
beta1[1]	14.90	6.205	7.28	26.66	1.18	10
beta1[2]	−0.51	0.549	−1.28	0.15	6.39	2
beta1[3]	0.02	0.130	−0.13	0.19	7.69	2
beta1[4]	−0.26	0.298	−0.70	0.10	1.39	6
beta1[5]	0.06	0.181	−0.16	0.34	2.18	3
beta1[6]	−0.28	0.253	−0.68	−0.00	3.03	3
Deviance	3246.72	211.948	2960.990	3557.981	3.28	2

Bayesian Weibulll distribution fitting on survival data

```
survweibMC(m1=5,n1=9,m2=1,n2=4,chains=2,iter=10,data=exdata1)
```

Listing 4.2: Output on Bayesian Weibulll distribution

	mu.vect	sd.vect	2.5%	97.5%	Rhat	n.eff
beta1[1]	7.41	2.15	4.08	9.64	2.46	3
beta1[2]	0.00	0.12	−0.12	0.22	1.08	10
beta1[3]	0.01	0.02	−0.02	0.04	0.93	10
beta1[4]	0.08	0.19	−0.18	0.34	4.16	2
beta1[5]	0.08	0.04	0.02	0.16	1.34	7
beta1[6]	−0.07	0.08	−0.22	0.06	0.96	10
deviance	3408.14	526.57	2906.97	4157.13	8.26	2

Bayesian Exponential distribution fitting by avoding death due to other causes

```
exdata2a=subset(exdata1,cmprsk!=2)
survexpMC(m1=5,n1=9,m2=1,n2=4,chains=2,iter=10,data=exdata2a)
```

Listing 4.3: Output on Bayesian Exponential distribution by adoving death due to other causes

	mu.vect	sd.vect	2.5%	97.5%	Rhat	n.eff
beta1[1]	10.67	3.13	4.92	14.41	1.03	10
beta1[2]	0.85	1.18	−0.41	2.24	13.34	2
beta1[3]	−0.21	0.20	−0.46	0.00	12.20	2
beta1[4]	0.40	0.50	−0.26	1.05	2.22	3
beta1[5]	−0.00	0.09	−0.18	0.11	1.25	7
beta1[6]	−0.42	0.29	−0.86	−0.03	3.95	2
deviance	2705.95	452.56	2199.24	3272.93	9.791	2

Bayesian Weibull distribution fitting by avoding death due to other causes

```
survweibMC(m1=5,n1=9,m2=1,n2=4,chains=2,iter=10,data=exdata2a)
```

Listing 4.4: Output on Bayesian Weibull distribution fitting by considering death due to other causes

```
            mu.vect sd.vect    2.5%   97.5%  Rhat n.eff
betal[1]      12.14    3.86    7.58   18.14  2.64     3
betal[2]       0.19    0.17   -0.07    0.52  1.15    10
betal[3]      -0.01    0.03   -0.07    0.03  1.14    10
betal[4]       0.03    0.25   -0.41    0.33  2.16     3
betal[5]      -0.00    0.07   -0.07    0.13  1.78     4
betal[6]      -0.24    0.10   -0.36   -0.04  1.06    10
deviance    2239.26   46.44 2184.95 2293.18 11.12     2
```

Bayesian Exponential distribution fitting by considering death due to other causes

```
exdata2b=subset(exdatal,cmprsk!=1)
survexpMC(m1=5,n1=9,m2=1,n2=4,chains=2,iter=10,data=exdata2b)
```

Listing 4.5: Output on Bayesian Exponential distribution fitting by considering death due to other causes

```
            mu.vect  sd.vect    2.5%   97.5% Rhat  n.eff
betal[1]       7.56  5.51      -3.19   12.26 2.65  3
betal[2]       0.20  0.59      -0.32    1.28 2.46  3
betal[3]      -0.04  0.13      -0.28    0.07 2.97  3
betal[4]       0.24  0.37      -0.22    0.74 5.36  2
betal[5]       0.07  0.16      -0.12    0.37 1.51  5
betal[6]      -0.18  0.10      -0.41   -0.09 1.44  6
deviance    1824.30  78.27   1712.52 1954.51 2.67  3
```

Bayesian Weibull distribution fitting by considering death due to other causes

```
survweibMC(m1=5,n1=9,m2=1,n2=4,chains=2,iter=10,data=exdata2b)
```

Listing 4.6: Output on Bayesian Exponential distribution

	mu.vect	sd.vect	2.5%	97.5%	Rhat	n.eff
beta1[1]	7.28	4.27	0.33	14.24	1.12	10
beta1[2]	0.29	0.28	−0.16	0.68	3.93	2
beta1[3]	−0.03	0.05	−0.11	0.05	5.02	2
beta1[4]	0.11	0.34	−0.38	0.59	1.56	5
beta1[5]	0.17	0.10	0.04	0.37	0.90	10
beta1[6]	−0.20	0.12	−0.34	0.04	2.48	3
deviance	1454.94	15.06	1438.79	1481.79	3.40	2

Define Data

```
library(SurvHiDim)
data(hnscc)
survintMC(7,11,Leftcensor="leftcensoring",OS="os",
Death="death",6,hnscc)
```

Listing 4.7: Output on Bayesian Exponential distribution

Variables	HR	LCL	UCL
GJB1	0.90	0.89	0.90
PPP1R9A	0.92	0.92	0.93
HPN	1.00	0.99	1.00
SLC4A4	0.97	0.96	0.97
HNF1B	0.89	0.88	0.90

Listing 4.8: hidimSurv outcomes on competing risk analysis

Variables	HR.x[LCL.x,UCL.x]	HR.y[LCL.y,UCL.y]
AFF3	0.88[0.82,0.95]	0.88[0.79,0.97]
B3GAT1	0.90[0.82,0.98]	0.88[0.78,0.99]
METTL7B	1.09[1.02,1.16]	1.09[1.00,1.18]
PGC	1.14[1.02,1.28]	1.18[1.02,1.36]
RBP4	1.07[1.02,1.13]	1.08[1.01,1.15]
ZMAT1	0.89[0.84,0.96]	0.88[0.80,0.96]

Chapter 5

Biomarker Thresholding in High Dimensional Data

5.1 Introduction

A biomarker is a biological feature that is presented to identify patients from the general population. Detection of the threshold limit value (TLV) of a biomarker is a complex procedure. Traditionally, the TLV is used to classify patients from the general population to improvize the therapeutic efficacy. The efficacy of a clinical trial improvizes due to a successful adaption of a biomarker thresholding [6, 7]. However, results obtained from subgroups cannot be generalized to all patients due to the lack of evidence for all. It helps conduct any subgroup analysis while there is evidence about the better performance of therapy of patients. Now the subgroup analysis for thresholding preparation can be formulated based on a specific data distribution of a biomarker. It is required to create the threshold to define a biomarker. It helps to predict the disease progression by looking dynamic of the biomarker [8]. It works with a small sample of patients to upgrade the therapeutic effect of the biomarker in a subgroup [6]. However, it becomes challenging while the prevalence of a biomarker becomes low in the general population. So persons having a high value of biomarkers can only be considered for a trial. Now the only way is to perform the targeted therapy among a group of individuals. The inclusion criteria and management criteria require to improve to achieve the targeted therapy [6]. Now, prior knowledge about biomarkers also helps to upgrade the stratification of the patient population by looking at the response rate. Now the development of the validating biomarker is a tedious process. It is generated through the strength of the existing evidence of the biomarker. The evidence source comes from the prior studies [9, 10]. Now the biomarker

DOI: 10.1201/9781003185598-5

can be classified as (I)Classified Biomarker, (II)Prognostics Biomarker, and (III)Predictive biomarker. But all biomarkers are not dynamic in change. If the biomarkers are dynamic, then it requires capturing the data as time-course data. It becomes more challenging to have subgroup-driven analysis for dynamic biomarkers.
The DNA biomarker is one of the statistic biomarkers that never changes over time. It is ethical and convenient to run any trial once biomarkers are settled down and established as prominent biomarkers. These are defined as prognostic or predictive biomarkers. Now any new therapeutic effect is demonstrated by a predictive biomarker. This predictive and prognostic biomarker can be explored and presented by treated and control arms with or without biomarkers. So this predictive biomarker can be established by comparison of treated and control components with or without the biomarker. So the duration of survival can be compared based on the presence of the biomarker. It helps to consider the biomarker and present it as a potential predictive biomarker. But follow-up duration becomes tedious to measure because this duration often requires maintaining the follow-up for several years. It is time taking procedure to establish any biomarker for cancer progression [7, 11]. The alternative of prolonged survival can be covered by progression-free survival or disease-free survival. Sometimes it can be formulated by response rate. Now, this response rate can be formulated as complete response, partial response, progressive disease, and stable disease [12]. However, the dynamic expression value of the prognostics marker changes its characteristics, not by treatment efficacy. For example, mutated Non-small cell lung cancer(NSCLC) patients are always better in comparison to non-mutated patients, whether they are treated with Epidermal growth factor receptor(EGFR) inhibitors or chemotherapy [13–16].

5.2 Statistical Methodology for Biomarker Thresholding

Now Bayesian hierarchical modeling is suitable to adopt to decide the specific therapy for patients [17]. It helps to determine the particular treatment by 'borrowing strength' about the biomarker as covariates from earlier treatment outcomes [17]. The application of the Bayesian hierarchical probit model is helpful in this context by subgroup-based clinical trial outcomes [18].

However, these come through adaptive clinical trials. It requires the rigorous development of statistical algorithms and design. But the progress of statistical methodology with high dimensional data for the biomarker is limited. This context requires classifying the patients based on the biomarker-based expression. Since several biomarkers combine role in disease progression, all biomarker needs to be defined as covariates.

So the statistical models are presented by the interaction of the covariates on therapeutic arms. The biomarker as covariates requires to give as a predictive biomarker as significant interaction test results [19]. Biomarker driven statistical model needs to be present for the predictive biomarker, not for the prognostics biomarker. Now the development of statistical methodology by two-stage sampling and type-I error is attempted [20]. It is helpful to consider the patient population by the subgroup by detection of treatment outcomes [20]. A simulation study for hypothesis testing of the global treatment effect with reference to the biomarker-positive subgroups can be performed. The correlation test between test statistics was achieved through the bivariate normal model [21]. There is also an attempt to check the multiplicity to control the false positive rate on composite hypothesis testing. A particular type of alpha spending function with a minimum p-value is considered to deal with multiplicity problem [22].

5.3 Thresholding for Repeatedly Measured Biomarker

Now the development of statistical methodology by two-stage sampling and type-I error is attempted [20]. It is helpful to consider the patients population by the subgroup by detection of treatment outcomes [20]. A simulation study for hypothesis testing of the global treatment effect with reference to the biomarker-positive subgroups can be performed. The correlation test between test statistics was achieved through the bivariate normal model [21]. There is also an attempt to check the multiplicity to control the false positive rate on composite hypothesis testing. A particular type of alpha spending function with a minimum p-value is considered to deal with multiplicity problem [22]. The prognostic biomarker changes over time during the study period. It requires measuring the prognostics biomarker thoroughly during the study.

Similarly, it is not easy to measure the prognostic biomarker value repeatedly throughout the study. But there is a recent trend to capture the time-course gene expression data [23, 24]. But the challenge with dimensional reduction is always faced in high dimensional time course data [25]. So it can be overcome by the gene set analysis [26]. It is helpful to consider gene set analysis as a powerful tool in time course data [27]. The gene expression identifies the potential biomarker by randomized clinical trial to examine the specific biomarker value to concentrate on particular biomarker value and control their threshold limit value (TLV) at the desired level to provide the best practical therapeutic effect. In oncology, the primary endpoint is to prolong the survival duration. The duration of survival is often called overall survival (OS). Now the information on prognostic biomarkers observed at different time points is another dimension to be concentrated. The merging area of time-course data with survival analysis becomes the ultimate choice to deal with a prognostic biomarker. The joint models provide the scope to extend the application of individual-specific repeatedly measured data together with OS. The repeatedly measured data may be a prognostic biomarker. The repeatedly measured correlated data are not analyzed jointly with OS and are prone to generate biased results due to avoiding the terminal event, which is competing for an event for the occurrence of repeatedly measured data [28]. Similarly, the conventional analysis with OS may lead to biased estimation if the repeatedly measured data are assumed as time-varying covariates or if it is ignored. The joint model for repeatedly measured data and terminal event data are required to be merged. There is an attempt through multiple failure times [29], and multiple longitudinal outcomes [30]. The repeatedly measured data can be merged with time-to-event data through the frailty model. In frailty model, the subject effects are assumed as frailty term [31, 32].There is a possibility to incorporate the non-parametric covariates in frailty model [33], disease-specific event process [34] and time-varying covariates [35, 36].There is an attempt with a Bayesian approach in joint frailty models [37]. It is possible to link the repeatedly measured data and death events via random effect. The marginal model in generalized linear mixed model structure is found suitable to deal with joint longitudinal and survival model [38]. The repeatedly measured trivariate data is proposed [39] and performed with cumulative intensities [40]. It is only performed for medical cost data [41]. The extension with latent structure through random effect models is attempted [42]. It is possible to obtain dynamic prediction of repeatedly measured data though jointly model on time-to-event and recurrent event [43, 44],trivariate models [42] and recurrent and terminal event for repeatedly measured data [45].The receiver operating characteristic (ROC) [46] and Brier score are found suitable to test the performance of predictive accuracy in the joint model for repeatedly measured and survival data [42, 44, 45]. Now the exploration of the prognostic marker may be helpful in detecting the biomechanical changes with novel therapy. It may lead to the detection of a new targeted agent for cancer treatment. This study is dedicated to the development of statistical methodology for the analysis of prognostic biomarker enrichment analysis.

5.4 Statistical Model

The cox model with time dependent covariates is defined as

$$h(t|x(t)) = h_0(t)\exp(x(t)^T\beta(t)) \tag{5.1}$$

The baseline hazard function defined as $h_0(t)$.The measurement of covariates at time t is defined as $x(t)$ and regression coefficient is $\beta(t)$. The covariates can be defined with different parts. The covariates may be time-independent or dependent. The prognostic biomarker is defined as time-dependent covariates. It is observed as a continuous variable. It is expected that the Prognostic biomarker will carry a time course trajectory over the follow-up period. The time-varying component helps to get an idea about covariates at baseline measurement. It is obvious that the regression parameter will change over time. The distributional assumption and test also play a crucial role in defining covariates as prognostics markers or not. However, the distributional assumption may be avoided through the application of the conditional model by available information at time point $t = s$.

$$h_s(t|x(s)) = h_{s,0}(t)\exp(x(s)^T\beta_s(t)) \text{ for} s \geq t \tag{5.2}$$

The relation between regression parameter $\beta_s(t))$ in the conditional model provides the estimates of regression in Cox model. It is assumed that the $x(s)$

gives the predicted value of $x(t)$. It is defined as

$$x(s)^T \beta_s(t) \approx E[X(t)^T \beta(t) | X(s) = x(s)] \tag{5.3}$$

or

$$\beta_s(t) \approx \frac{cov(X(s), X(t))}{\text{var}(X(s))} \beta(t) \tag{5.4}$$

Further, the estimates of $\beta_s(t)$ stands with range $[s_1, s_L]$ having conditional baseline hazard model. The covariance can be processed through $X(t)$ by

$$\beta_s(t) = \beta(s) \times \lambda(t - s) \tag{5.5}$$

The function $\lambda(t - s)$ is monotonically decreasing with $\lambda(0) = 1$ to $\lambda(\infty) = 0$. It is suitable to consider $\lambda(t - s) = exp(-v(t - s))$ as non-linear dependence with v [47]. The simplest handling procedure could be assumed with the range of survival prediction by s to $s + w$. The model is split with

$$h_{s,0}(t) = h_0(t)\text{exp}(\theta(s)) \tag{5.6}$$

where $\theta(s)$ is formulated with basis function $g_j(s)$ as

$$g_j(s_1) = 0\text{and}\theta(s) = \sum_{j=1}^{m_k} \eta_j g_j(s), \tag{5.7}$$

where the estimate is defined as

$$\hat{h}_{s,0}(t_i) = \frac{1}{\sum_{j \in R(t_i)} \text{exp}(x_j(s)^T \beta(s))} \text{for } s \le t_i \le s + w \tag{5.8}$$

The term s predicted through the regression coefficient of $\beta(s)$ through the base value of $x(s)$. The dataset needs to be through the combination of conditional point s and event time t. It is important to capture the observations of prognostic biomarkers continuously. It is possible to capture through the time-dependent covariate. However, the last observation carries forward method can be considered to approximate the unavailable information of a variable for the patients. The subgroup-defined study can recruit the patients with a high level of TLV and proceed to predict their survival time point based on baseline characteristics of different covariates. The result with time-dependent covariates with the Cox model is performed on thyroids data. This procedure is simple to get the prediction for a subgroup-oriented study. Patients who have more than TLV of a prognostic biomarker are recruited for the study. Now data will consider patients with more than TLV and predict their survival time point based on baseline characteristics of the time-dependent covariates.

The Cox proportional hazards(CPH) is the widely used method in survival analysis. There is a similarity between the multilevel model and random effect CPH modeling. In both cases, the cluster-wise homogeneity is assumed through the random effect component. The CPH with random effect is also known as the "frailty model" However, it is better to define them as CPH mixed effect [48], as it provides scope to incorporate multiple random effects. The frailty model does not have the option to incorporate specific components as a random effect. In this work, the CPH mixed effect model is defined as

$$h_i(t) = h_0(t)\text{exp}(X_i\beta + \alpha_j) \tag{5.9}$$

The random component associated with j^{th} cluster is defined as α_j. Further, it can be separated into

$$h_i(t) = h_0(t)\text{exp}(\alpha_j)\text{exp}(X_i\beta) \tag{5.10}$$

The shared frailty term is suitable to address the CPH with a mixed effect model. It is possible to explore gamma, log-normal as a distributional assumption for random-effect terms (better known as shared frailty term) [49–51]. The presence of two or more levels of clustering is defined as a "nested frailty model" for survival analysis [52]. However, the "nested frailty model" is limited to one level with two units [53]. The Cox Proportional Hazards with Mixed Effects model is suitable enough to handle a determined subgroup study. The selected patients with a specific threshold can be further explored with different covariates and their influence on RFS or OS. In this study, the Thyroid is considered a covariate among recruited patients. The survival status and duration of survival of ith patients is defined with δ_i and T_i respectively. The true survival duration is defined as T_i^* and censoring time as C_i. The value of δ_i becomes 1 when $T_i^* \leq C_i$ and 0 otherwise. The repeatedly measured observation of the ith patients j^{th} observation is defined as $y_{ij}, i = 1, 2, .., n$ and $j = 1, 2, ..l$. The response variable is defined with linear mixed effect model by

$$y_{ij} = \mu + x_{ij}^T\beta + z_{ij}^T b_i \tag{5.11}$$

The longitudinal response is defined as y_{ij}. The fixed effect model is defined as x_{ij} and b_i as random effect. The time-dependent design vector is assigned with z_{ij}. It is assumed that b_i follows the $N(0, \sigma^2)$ distribution. The baseline hazard is defined as $h_0(t)$. The hazard function is defined as

$$h_i(t|H_{ij}, w_{ij}) = \lim_{\Delta t \to 0} \frac{1}{\Delta t} P\{t \leq T_i^* < t + \Delta t | T_i^* \geq t, H_{ij}, w_{ij}\} \tag{5.12}$$

$$h_i(t|H_{ij}, w_{ij}) = h_0(t)\exp[\gamma^T w_{ij} + f\{H_{ij}, b_i, \alpha\}], t > 0 \tag{5.13}$$

The time-dependent covariates is defined as w_{ij}. The corresponding coefficient is γ. The baseline hazard function is defined as $h_0(.)$. The hazard function through consideration of longitudinal process is defined as $H_{ij} = \{\eta_i(s), 0 \leq s < t\}$. The term α is used to make a bridge between prognostic biomarker upto j^{th} observation and hazard of an event at j^{th} time point. In this study, the model is performed through Bayesian approach by Markov chain Monte Carlo (MCMC) algorithms. The random effect component in longitudinal and survival model is assumed as independent in nature.The model is defined as

$$p(y_i, T_i, \delta_i, \theta) = p(y_i|b_i, \theta)p(T_i, \delta_i|b_i, \theta) \tag{5.14}$$

$$p(y_i|b_i, \theta) = \prod_j p(y_{ij}|b_i, \theta) \tag{5.15}$$

The posterior distribution is formed as

$$p(\theta, b) \propto \prod_{i=1}^{n} \prod_{j=1}^{l} p(y_{ij}|b_i, \theta)p(T_i, \delta_i|b_i, \theta)p(b_i|\theta)p(\theta) \tag{5.16}$$

5.5 Repeteadly Measured Biomarker Thresholding

Generating repeatedly measured biomarker data is a difficult task in therapeutic effect comparison due to ethical and financial constraints. So help of a simulation study can be taken to improvize the existing statistical methodology. The values of the prognostic biomarker can be generated for different time points as simulation. The skewed distribution is considered to generate the prognostic biomarker data seen in lung adenocarcinoma patients‘ data. It is documented that circulating endothelial cells (CEC) as a prognostic biomarker for various primary and metastatic cancer [54–56]. The higher rates of CEC penetrate the tumor cells, and tumor progression [57–59]. It is confirmed that dynamic changes of CEC are promising predictors of the response to treat anti-angiogenic agents [60, 61]. It is not feasible always to generate the time-course measurement of CEC due to logical and ethical issues [62]. A total of 120 patient data is generated by simulation and similar to the real situation. The CEC values are obtained in a time-dependent manner. The normal distribution assumption with mean 114 and SD 15 is considered to generate the distribution. As a choice of different treatments, a total of four choices, i.e., dose 1, dose 2, dose 3, and dose 4, are prepared. The best performing dose is defined as the dose that can control/reduce the CEC measurement. The CEC value of the different points is generated with a mean of 124, 128 and 135, respectively. The SD value 15 is considered to generate CEC values for different time points. The values of tumor size were obtained with mean values of $4.0, 3.4, 3.2$ with SD value 2 for time points. The time points are defined as t_1, t_2 and t_3 respectively. The duration of relapse-free survival is defined as the least value between uncensored and censored survival duration. The relapse-free survival is generated with a hazard rate of 0.0003 with the presence of 49% censored observations. The graphical representation of generated data through a spaghetti plot is provided in Figure 5.3.

Now a Bayesian framework is created to solve the problem. We started our simulation study with 120 patients. The first post-baseline visit is measured with patients with the controlled value of CEC. It is observed that a total of $n = 30$ patients treated with dose 1 and $n = 13$ are observed with a lower value of CEC in the first visit in comparison to baseline. In subsequent visits, the patients observed with lower/controlled values of CEC for dose 1 were 16 and 18, respectively. The total number of patients treated with dose 1 for each visit is 30. Now if $n = 13$ or $n = 16$ or $n = 18$ there is at least 0.95 posterior probability of superiority to claim any statistical benefits. It is generated as $X_1 > 13, X_2 > 16$, and $X_3 > 18$, respectively. The variables X_1, X_2, X_3, are presented to formulate the doses and their success rate. The sample size is calculated with a Type I error.

$$1 - \sum_{i=0}^{12} \sum_{j=0}^{15} \sum_{k=0}^{17} \Pr(X_1 = i)\Pr(X_2 = j - i)\Pr(X_3 = k - j) \tag{5.17}$$

where X_1, X_2, and X_3 are assumed to follow binomial distribution with success probability p. The output of this simulation study is detailed in Table 5.1. It becomes feasible to obtain power values for the different scenarios. The simulation is performed with 100000 iterations. It shows while the hypothesized value of p is 0.70, then the power of this study is observed at 0.534. The mean sample size is observed with 15.09 with a probability of 0.419 to make a decision about whether dose 1 is successful with patient sample 13 or not. Similarly, other p-values listed in Table 5.1 and their corresponding power can be adapted for the suitable decision about the performance of a repeatedly measured prognostic biomarker during the study. A similar procedure can be performed through dose 2, dose 3, and dose 4, respectively. The R code is given in annexure-I.

5.6 Biomarkar Thresholding Determination

Hypothyroidism occurs when the thyroid gland fails to produce enough T4, and the body fails to convert into the required level of T4 and T3 in peripheral tissues, or the nervous system does not work properly to stimulate the thyroid gland [62]. It creates complications due to the head, and neck irradiation [63].

The level of complication due to hypothyroidism may vary from age, site of the primary tumor, nodal stage, fractionation schedule, concurrent chemotherapy, and duration of follow-up. But it is unknown about the contribution of hypothyroidism toward tumor size suppression and the effect of a prolonged relapse-free period. The measurement of hypothyroidism value was obtained from a phase 3 randomized open-label prospective study to compare the weekly versus three-weekly cisplatin doses. A total of 300 patients were recruited in this study to assess the non-inferiority of cisplatin 30 mg/m^2 given weekly to cisplatin 100 mg/m^2 given 3-weekly, both administered concurrently with curative intent radiotherapy in patients with Locally Advanced Squamous cell Carcinoma (LAHNSCC). The primary endpoint was locoregional control.

Locoregional failure is defined as the recurrence of a tumor in a local or regional area. A total of 150 patients with cisplatin 30 mg/m^2 arm and 150 patients in cisplatin 100 mg/m^2 given 3-weekly are included. It is estimated that the cumulative 2-year locoregional relapse rate (LRR) was 38.67% in the weekly arm and 24.67% in the 3-weekly arm, leading to an absolute difference of 14% (95%CI, 3.59-24.41), p = 0.014. The locoregional control (LRC) was defined as the time from randomization to tumor recurrence in the original tumor bed, development of cervical lymph nodal metastases, or cutaneous nodules in the head and neck region after treatment.

At each visit, the complete history and relevant physical examination were performed along with blood parameters were tested inclusive of serum T_3, T_4, and thyroid-stimulating hormone (TSH) measured through thyroid function test (TFT). It gives the scope to get the measurement with the time course pattern. The LRC status and duration of an occurrence of LRC provide the scope to explore the time-to-event data analysis. Combinedly, the data with repeatedly observed Hypothyroidism and LRC make a sense to work with a joint longitudinal and survival model.

The repeatedly measured TSH value at 3,6,9,12 and 24 months are explored through the Cox PH model with LRC. There are some missing observations, and those are imputed through Multivariate Imputation via Chained Equations (MICE) package 2.46.0 for intermediate missing observations in R. One combined value of TSH for each patient is generated through six-time points weighted mean average for patients followed/censored till 24 months. Further, the cut-off value of the TSH level is decided through the application of the CPH model between the TSH level and LRC measurement. The cut-off value of TSH is decided on weighted mean averages. The P-chain diagram is considered to detect the cut-off value of TSH [64]. The least P-value is observed with a combined TSH level of < 40 or > 40 on LRC id detailed in Figure 5.2.

The threshold determination of biomarkers by the time-dependent covariates in the Cox Model and Cox proportional hazards with mixed effects are presented separately. Results obtained with separate longitudinal and survival analyses are given in Table 5.2.

The point estimate of the regression estimates is presented. The linear mixed effect model is helpful in obtaining the chemotherapeutic effect data and comparison. The linear mixed effect is obtained by the "lme" function from "nlme" packages in R. In this example, the therapeutic effect of weekly or three weekly chemotherapy is compared to the time of thyroid biomarker changes.

The regression coefficient for Trt(weekly) was obtained with −0.59 in Table 5.4. It shows that the level of Thyroid in the weekly arms was less in comparison to the three weekly arms. The survival analysis result is now presented in Table 5.2. Directly the patients with higher levels of the Thyroid are observed with more hazards. Patients treated with weekly arms received more benefits than three weekly arms.

A higher level of Thyroid inclines the hazard ratio in comparison to a low thyroid level with a regression coefficient of 0.02. This p-value chain helps to determine the TLV of the Thyroid as a biomarker. The TLV can be determined by creating the optimum value of Thyroid between < 40 and ≥ 40 of TLV. It can be obtained by p-value as given in Figure 5.2. The etwork plot using hdNetwork function using R given in Figure 5.4.

The subgroup-defined study generates the inference of novel treatment effects in a simplified and prompt manner. The challenge is to make inferences in the presence of limited power to assess the treatment effect. The advantage of determining the biomarker thresholding is that it ignores the introduction of bias toward patient selection. The inflation of the Type I error rate also protected toward detection of predictive biomarker signature.

We can analyze the data from non-inferiority randomized trial data investigating the therapeutic effect on thyroid levels. The repeatedly measured data provides the scope to accumulate the overall treatment effect through time-to-event in a longitudinal manner. The estimates of the hazard ratio were obtained with 1.01. The survival function coxph and survfit compatible with R can handle the time-varying covariates [65]. However, it is a little bit tricky to prepare the data properly. The fold function is suitable for dealing with data preparation. [66]. The "sursplit" function is useful to handle fold function generated data. The timereg package available in R is another useful function to carry the time-dependent covariates in the Cox model [67].

The comprehensive literature about time-dependent covariates in the Cox model is well-documented [68]. The hazard ratio was estimated at 1.003 through CPH with the mixed effect model. The estimate of the parameters observed through the longitudinal model is 0.02. Various estimates for hazard ratios could be obtained through different distributional assumptions. The flexibility of different distributional assumptions lies with the consideration of shared frailty terms. The gamma distribution and log-normal distribution are the best candidates under shared frailty terms. The detailed about random effect components and their distributional assumptions are well-documented [49–51].

In cancer clinical trials, joint modeling is generally avoided due to the involvement of cost for several and repeated investigations. Repeated investigations are required to closely monitor the disease progression with changes in the different markers. However, it is not always feasible. The majority of literature on longitudinal modeling is dedicated to the development of the model, not to studying design. The longitudinal model in cancer clinical research cannot be performed without survival modeling. Therefore, the joint longitudinal and survival model is an important area in cancer clinical research. In oncology research, the joint modeling of health quality of life data and time-to-event are elaborated separately [69–72]. The joint modeling is also elaborated in other applications as well [73–75]. The comprehensive review of joint longitudinal and survival modeling can be observed with [76]. The future of successive therapeutic effects in cancer research is personalized medicine. This subgroup-defined study design ensures the application of personalized medicine in cancer research for the selective patient with a specified dose. Therefore, it is important to prepare the study to accommodate longitudinal and survival processes to detect the therapeutic effect on the subgroup-defined study.

The joint model is now a well-established method for repeatedly measured and event-time data. There is not an attempt toward the development of biomarker enrichment analysis through joint longitudinal and survival modeling. It is possible to generate unbiased estimates in comparison to univariate analysis. Methods adopted in this work provide reliable estimates in simulated data. Each function is created for better compatibility in the models. The graphical representation of predictions may help compare the therapeutic effect. This work will lead to dynamic predictions for the joint modeling with a prognostic biomarker with a terminal and recurrent event. Longitudinal data distribution of all patients are presented in Figure 5.3.

Algorothm to create biomarker response rate

```
response1=function(SP, numberofsim ,PC,HP,nC)
v =logical(numberofsim)
ss = numeric(numberofsim)
NI = nC
for (i in 2:length(nC))
NI[i] = nC[i] - nC[i-1]

for (i in 1:numberofsim)
x=rbinom(length(nC),NI,SP)
x =c(x[1],x[1]+x[2],x[1]+x[2]+x[3])
ProbSup = 1 - pbeta(HP,1+x,1+nC-x)
v[i] =any(ProbSup > PC )
ss[i] = min(nC[ProbSup > PC],max(nC))

response= c(length(v[v])/numberofsim,mean(ss),
sqrt(var(ss)),table(ss)/numberofsim)
names(response) = c('Pr(v)','MeanSS','ss',as.character(nC))
response

numberofsim =1000000
PC= 0.96
HP= 0.50
SP= 0.7
nC= c(50,75,100)
response= response1(SP,numberofsim,PC,HP,nC)
```

Listing 5.1: Output with Posterior mean and SD

```
Pr(v)    MeanSS    SD       50      75      100
0.99     54.38     11.58    0.85    0.10    0.03
```

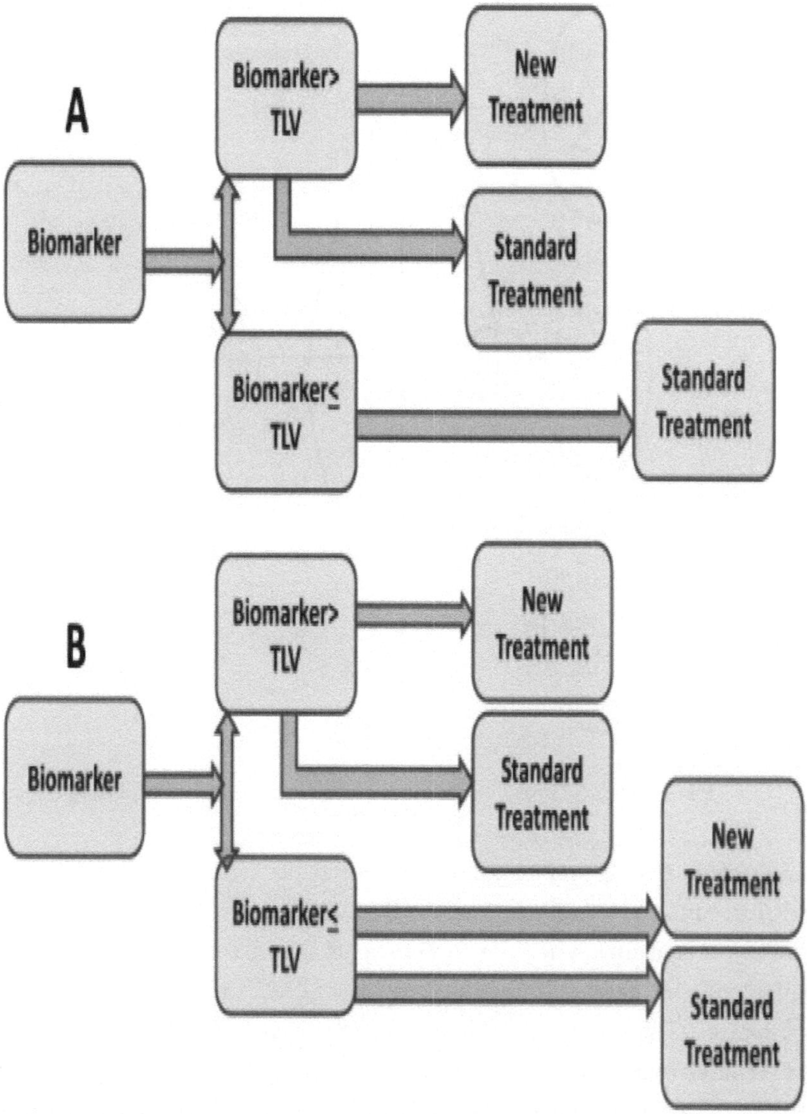

FIGURE 5.1: Biomarker-driven subgroup study to decide TLV (A).| Biomarker-driven subgroup study to decide TLV of having more than or less than TLV (B).

5.7 Illustration Using R

The working package on survival analysis and high dimensional data analysis are available in the "SurvHiDim" package. Anyone can download this

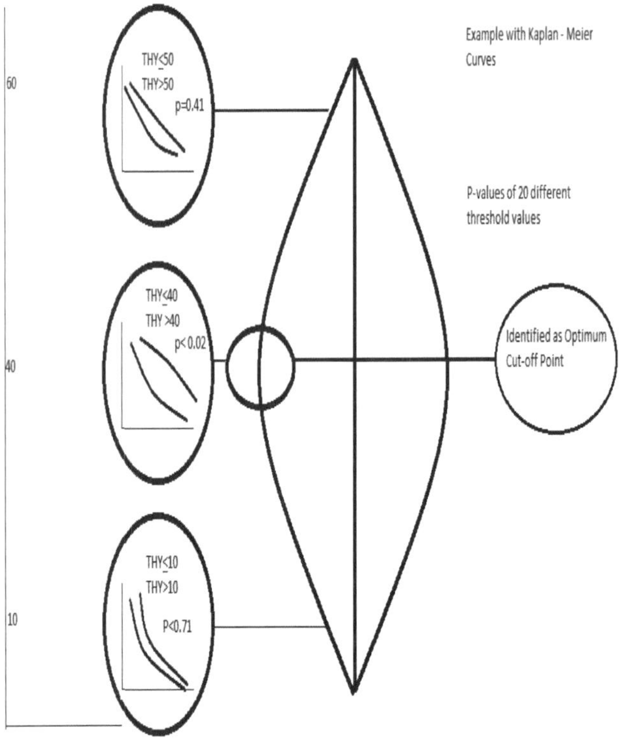

FIGURE 5.2: P-Chain derived for cut-off value of TSH

package from CRAN and install it into their R platform. Here, we present the "'hdClust" function available in "SurvHiDim" package. The dataset used for illustration is available in "'SurvHiDim" package. A total of 105 columns are available in this dataset. In addition, a total of 565 rows are also available in this dataset. The rows define the number of patients available in this dataset, and columns stand with high-dimensional data. A greater number of dimensions shows a high number of dimensions. Sometimes, it requires creating the network plot for high-dimensional data to list the number of variables. The "hdClust" function first creates a new column "Status" in the input data set and assigns values 0, 1, and 2 to each row. It assigns 0 (for progression = 1 & death(event) = 0) is or (when progression = 0 & death(event) = 0). Similarly, consider 1 (for progression = 1 & death = 1), whereas assigns 2 (for progression = 0 and death = 1).

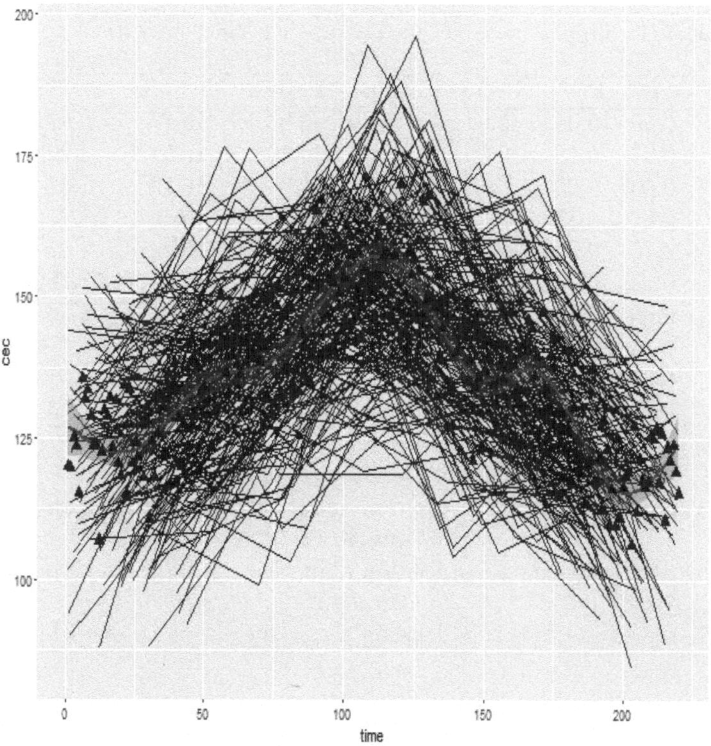

FIGURE 5.3: Longitudinal data distribution of all patients are presented.

Internally, it creates two data sets and one data set named 'deathdata', which includes subjects with status 0 and 1 and applies Cox PH on it. Another data is named 'compdata', which includes subjects with status 0 and 2, then applies Cox PH after substituting two by 1. Then it filters out study variables having P-value < sig-level(significance level taken as input from the user) from both subset data. Secondly, it merges the common significant variables from both data and creates a new data frame that contains columns, "ID", "OS", "Death", "PFS", "Prog", "Status" and observations of standard significant variables (which are supposed to be leading to death given they leads to the progression of cancer as well as accounts for competing risks). Further, it lists the common variables names and corresponds in results in .csv format by default in current working directory.

The "hdNetwork" function available in "SurvHiDim" package is useful to creates a new column "Status" in the input data set and assigns values 0, 1,

TABLE 5.1: Simulated table on Therapeutic Success with a specific dose level.

p	P(V)	Pr(13)	Pr(15)	Pr(17)	Mean SS	SD SS
0.50	0.06	0.04	0.02	0.93	16.77	0.87
0.55	0.13	0.09	0.04	0.86	16.54	1.20
0.60	0.23	0.16	0.06	0.76	16.19	1.52
0.65	0.37	0.27	0.09	0.62	15.69	1.77
0.70	0.53	0.41	0.11	0.46	15.09	1.87
0.75	0.70	0.58	0.11	0.29	14.42	1.78
0.80	0.84	0.74	0.09	0.15	13.81	1.48
0.85	0.94	0.88	0.06	0.05	13.35	1.01
0.90	0.98	0.96	0.02	0.00	13.09	0.51
0.95	0.99	0.99	0.00	0.00	13.00	0.13

P(V) = power of the study.
Mean SS = Mean Sample size.
Pr(13) represents to take a decision about success of dose1 with patients sample 13.
Pr(15) represents to take a decision about success of dose2 with patients sample 15.
Pr(17) represents to take a decision about success of dose3 with patients sample 17.

2 to each rows. It assigns 0 (for progression = 1 & death(event) = 0) is or (when progression = 0 & death(event) = 0. It assigns 1 (for progression = 1 & death = 1, whereas assigns 2 (for progression = 0 and death = 1).

Further, it creates two data sets and one data set named "deathdata", which includes subjects with status 0 and 1 and applies Cox PH on it. Another data is named "compdata", which includes subjects with status 0 and 2, then applies Cox PH after substituting 2 by 1. Then it filters out study variables having P-value < sig level (significance level taken as input from the user) from both subset data. Secondly, it merges the common significant variables from both data and creates a new data frame which contains columns, "ID","OS","Death","PFS", "Prog","Status" and observations of common significant variables (which are supposed to be leading to death given they leads

TABLE 5.2: Time-dependent covariates with Cox Model.

	exp(coef)	se(coef)	z	p	(LCL,UCL)
Thyroid	1.011	0.003	3.45	0.00	(1.00,1.01)

The Likelihood ratio test value is estimated as 10.11.

TABLE 5.3: Cox proportional hazards with Mixed Effects fit by maximum likelihood.

Parameter	Chisq	df	p	AIC	BIC
Integrated loglik	200.42	2.00	0	196.42	192.13
Penalized loglik	312.01	38.92	0	234.16	150.75
		exp(coef)	se(coef)	z	p
Thyroid		1.003	0.005	0.64	0.52

to the progression of cancer as well as accounts for competing risks). Further, it lists the common variables names and corresponds in results in .csv format by default in user's current working directory.

Perform the hdClust function using R

```
library(SurvHiDim)
data(hnscc)
hdClust(7,105,0.05,2,ID="id",OS="os",
Death="death",PFS="pfs",Prog="prog",hnscc)
```

Here paramters called from 7th to 105th column.

TABLE 5.4: Estimates of the parameters observed through joint longitudinal-survival models.

Survival Process					
Parameter	Estimate	SE	P-value		
Intercept	18.36	3.66	0.000		
TRT(weekly)	-.59	5.31	0.91		
obstime	0.00	0.00	0.59		
TRT weekly:obstime	0.00	0.01	0.76		
Longitudinal Process					
Parameter	Estimate	SE	P-value	Variance	
TRT(Weekly)	0.41	0.60	0.49	Random	18.22
Thyroid	0.02	0.01	0.05	Residual	22.76

Results obtained on hdClust function using R

K-means clustering with 2 clusters of sizes 3, 3
Cluster means:
Gene MeansForDeath MeansForAlive
1 5 -2.86 -2.78
2 2 -2.82 -2.89
Clustering vector: 2 2 2 1 1 1
Within cluster sum of squares by cluster:
2.589085 2.248709

Creates a network plot of high dimensional variables and lists those variables

```
data(hnscc)
hdNetwork(7,105,0.05,0.2,ID="id",OS="os",
Death="death"PFS="pfs"Prog="prog",hnscc)
```

Network matrix created by hdNetwork function using R

	AFF3	B3GAT1	METTL7B	PGC	RBP4	ZMAT1
AFF3	1.00	0.40	0.00	0	0.29	0.53
B3GAT1	0.40	1.00	0.00	0	0.00	0.46
METTL7B	0.00	0.00	1.00	0	0.36	0.00
PGC	0.00	0.00	0.00	0	0.00	0.00
RBP4	0.29	0.00	0.36	0	1.00	0.21

5.8 Data Illustration

In R using **rpart** package for high-dimensional data, then there are no covariates found which are highly significant on scaling of p-values (<=0.05 and 0.01) because of adjustment of multiple tests [77]. The scaling of p-values which is usually useful (for example 0.01, 0.05) is remain emphasized without adjustment [78,79]. For a large number of covariates, this is too time-consuming also,

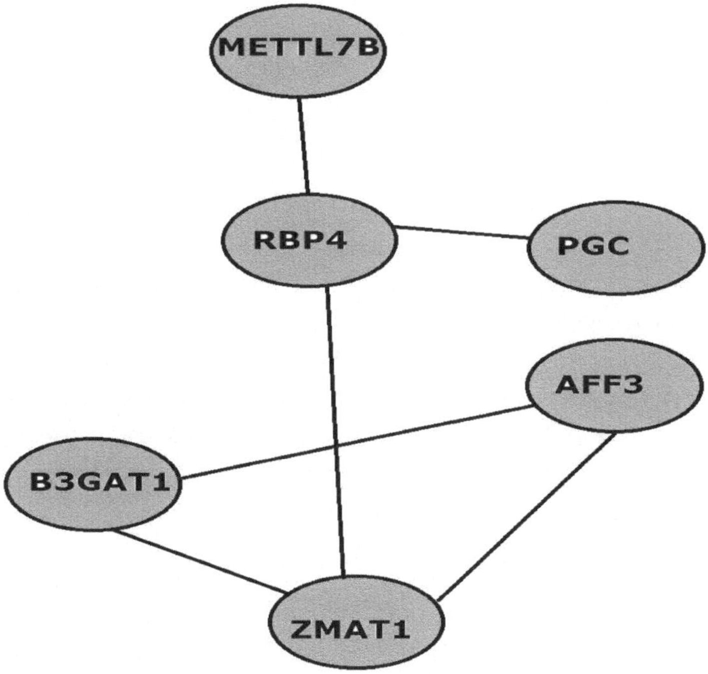

FIGURE 5.4: Network plot using hdNetwork function using R

and hence some researchers have suggested the univariate score test for the selection of significant covariates [80]. Multiple biomarkers in combined form can improve the accuracy of classification than using a single biomarker [81].

It motivated us to combine all biomarkers together at the stage of classification. The classification technique is provided for both multivariable and univariate covariates using Bayesian. In the time-to-event data, it is necessary to identify the variation in the gene expression (covariates or predictor variables in the model) based on the event's occurrence and classify covariates accordingly.

The coding algorithm is prepared to classify the Genom covariates and their survival estimates. The overall survival for breast cancer patients using clinical data and continuous-valued genomic data is presented. The flowchart for the selection of the studied six datasets meeting the criteria is illustrated in Figure 5.5. These 6 independent datasets (GSE3494, GSE7390, GSE16446,

GSE20685, GSE20711, and GSE48390) together include 1037 unique breast cancer patients. Dataset GSE3494 (236) was collected in Uppsala (Sweden) from 1 January 1987 to 31 December 1989 and published on the PNAS website, and approved by Karolinska Institute, Stockholm, Sweden [82]. Gene expression data GSE7390 (198) is taken from [83]. The diagnosis duration of patients was 1980 to 1998, and validation of data was done at all six clinical centers where data is seen [83] by two independent auditors. Dataset GSE16446 (107) is taken from [20], in which microarray data profiles were assessed using pre-epirubicin biopsies, and validation was done with MDACC 2003-0321 neoadjuvant trials. Dataset GSE20685 (327) is accessed from [84] with a diagnosis period from 1991 to 2004 at the Koo Foundation Sun-Yat-Sen Cancer Center (KFSYSCC). Dataset GSE20711 (88) is taken from [85] and available at Infinium Methylation Platform to find the profile of breast tumors at single-CpG resolution. Dataset GSE48390 (81) is taken from [86], and the diagnosis duration was January 2007 to December 2010 for breast cancer patients from Taiwan. These six datasets are taken from different sources on homogenous breast cancer patients taken from microarray experiments and were performed on a similar platform and direction. These datasets can be downloaded from https://www.ncbi.nlm.nih.gov/geo/ using a data access number.

The primary objective of the data is aligned with the overall survival(OS) of breast cancer patients. All these 6 data are combined in a single data of 1037 patients to be meta-analyzed. Here some conversions have been made, like the unit of survival time (overall survival) in all six independent data, converted into a single unit "day" whether it was in the day, month or year.

Overall survival, time to relapse, last follow-up, follow-up duration, and survival time all have been written in a single notation OS. For the event status, one indicates the occurrence of certain death status, and zero is for living or censored-biological accession ID. Reference of biomarkers into genomic data are $216836 - s -$ at for HER2, $205225 -$ at for ER, 219197 – s – at for PR, and $203497 -$ at for ERBB2.

Breast cancer varies from one geographical location to another geographical location [87]. Studied microarray data comes from different geographical locations while all have the same scenarios and same study type. So, we should expect a complete integrated study, which will have representation from other sites to give an overall survival estimate about the actual scenario of different biomarkers HER2, ER, PR, and ERBB2. We made a classification study on it. We represented our work through CART analysis toward detecting the threshold values for biomarkers HER2, ER, PR, and ERBB2. From CART analysis, we are getting the cut-off value of all four biomarkers (predictor variable) HER2, ER, PR, and ERBB2 on overall survival for better and faster breast cancer treatment.

Further studies are made on classified data with Cox proportional hazard (CPH) model for estimating the estimates (Hazard Ratio, Confidence Interval, and p-value) using the maximum likelihood method and the Bayesian approach using open-source R software. The CPH model is used to know the effects of biomarkers (covariates) on the occurrence of an event (death) and to make predictions on OS. All these estimated pieces of information are kept in Tables.

We performed a similar strategy to validate CART analysis on training data. As an arbitrary choice of 85% data, we selected from the entire dataset of 1037 sample size as training data for CART analysis toward validation of our methodology. We used the cut-off value obtained from the training data to classify the testing data of sample size 157(15%). We can also generate a Brier score of testing data to check the accuracy of the methodology in terms of risk or mean squared error. The Brier score can be obtained with R software using packages "DescTools".

5.9 Classification and Regression Tree Analysis in Biomarker Thresholding

The Classification and Regression Tree Analysis(CART) is helpful in obtaining the threshold value of a biomarker using R. CART is preferable as a data-driven method. It does not require making any hypothesis for the distribution of the predictor variable. It is suitable for highly skewed or multi-model data (numerical), and it is an optimal tree-building technique [88]. The model is used to perform CART analysis in the regression model, in which covariates are considered biomarkers, and the outcome was an event (death). Model trees were preferred to create the classification of predictor variables.

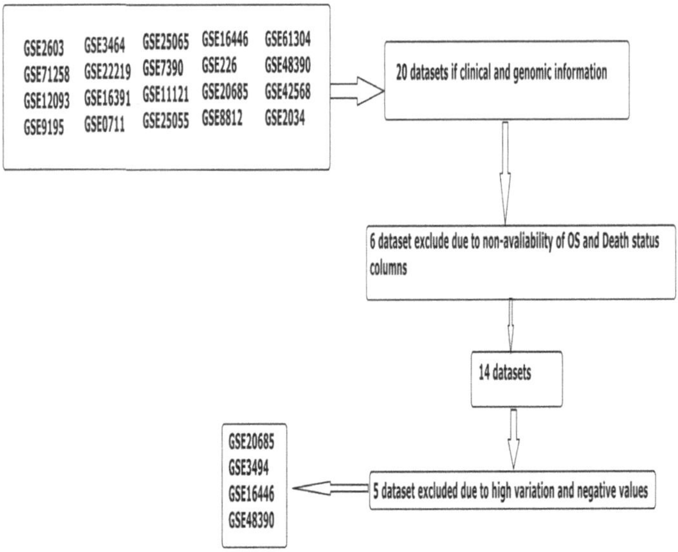

FIGURE 5.5: Flowchart for selection of data to make study on overall survival.

CART is chosen because we want to have the binary classification for our continuous predictor variables in the data. After performing CART analysis, we can classify the data. Parameters can be estimated using the CPH model. Cox regression analysis (CRA) is based on the CPH model. The robust nature of the CPH model permits us to find survival function estimates for the unknown parameter using the correct parametric model. With it, we can obtain the MLE of unknown parameters with asymptotic deliberation [89].

Inferences of MLE depend on the likelihood of the data. The Cox regression model is most commonly used in regression models, as it does not require any hypothesis or assumptions regarding the nature or shape of the distribution [90]. CPH model is executed in this study to make a comparison among biomarkers HER2, ER, PR, and ERBB2 on OS. The performance of these biomarkers is analyzed by Hazard Ratio (HR), Confidence Interval (CI), and p-value. Hazard function H(t) is expressed by

$$H(t) = h_0(t) exp^{\sum_{i=1}^{p} \beta x_i} \tag{5.18}$$

where t is survival time, $H(T)$ is the hazard function for p covariates $X_1, X_2, \ldots\ldots, X_p$ $h_0(t)$ indicates baseline hazards at time t and coefficients that measures the influence of predictors is denoted by β. The Hazard ratio for two covariate vectors Y_1 and Y_2 can be obtained by the ratio of hazard rates of respective covariates. It is expressed as

$$HR = \frac{h_{y_1}(t)}{h_{y_2}(t)} = \frac{h_0(t)e^{y_1\beta}}{h_0(t)e^{y_2\beta}} = \frac{e^{y_1\beta}}{e^{y_2\beta}} \tag{5.19}$$

Hazard ratio quantifies the degree of difference between the groups. The event's occurrence (i.e., risk) increases if $HR > 1$ by (HR $\times$ 100)% , the event's occurrence (i.e., risk) decreases if $HR < 1$ by (1 $-$ HR $\times$ 100)% , and if $HR = 1$ then it describes lack of association.

In addition, the receiver operating characteristic (ROC) curve is prepared to measure the accuracy of created classification. Predictions obtained by Bayesian survival analysis (BSA) have very little bias and standard error than the Cox regression analysis (CRA), whether the sample size is large or small [91]. BSA uses the Bayesian method of estimation, in which inferences are exact. The Bayesian approach for estimation is based on prior information and new knowledge from the experimental (observed) data [92]. Bayesian analysis is mainly used for informative and non-informative prior.

Informative prior is based on earlier experience, studies, or expert views, which are not conquered by likelihood, used to find posterior distribution. In some cases, when prior information is unavailable, non-informative prior density is used, which is conquered by the likelihood function. It has the minimal influence on posterior distribution [92, 93], and improper posterior can be obtained with it [94].

We have used two ways of estimation, MLE, and Bayesian, to estimate the parameters. Prior information plays a vital role in Bayesian survival analysis. Here, regression coefficients are the parameters in the CPH model, in which prior distribution is taken as normal. The Markov Chain Monte Carlo (MCMC) method used the posterior density parameter. For estimating the parameters using the Bayesian approach, we have used the function "survMC" from the package "SurviMChd" with 10000 iterations, which provides survival estimates using the CPH model with MCMC.

MCMC method is used in multi-dimensional data for the solution or explanation of sampling distribution. Monte Carlo is used for sampling (randomly) a probability distribution to estimate the parameters, whereas the Markov chain is a systematic method for generating the sequence of random variables, in which the current variable is generated using a primary variable.

Using MCMC, many samples can be generated from posterior density, and it is used to estimate the study of characteristics by taking its expectation. MCMC has many applications in statistics. For example, the Gibbs sampler is a powerful simulation technique, and it is a method of MCMC.

Risk prediction of the methodology is one of the key tasks. Brier score (BS) is used in this article to validate the accuracy of the classification model using its testing data. This illustrates validation in terms of mean squared error between predicted probabilities and expected values. The expression to calculate the Brier score is

$$BS = \frac{1}{n}\sum_{j=1}^{n}(p_j - o_j)^2 \tag{5.20}$$

Where p_j is predicted probability (takes a value between 0 and 1) and O_j denotes actual outcome (takes a value of 0 or 1), and the Brier score measures the precision of probabilistic estimation. As much as it gives lower values, it means a better prediction is obtained.

Recently, boosting has been used as an optimization technique to estimate the potentially high-dimensional models, whether additive or linear [95–97]. There is a 'mboost' (model-based boosting) package in R software for statistical modeling to minimize the empirical risk function $\frac{1}{n}\sum_{i=1}^{n}\rho(Y_i, f(X_i))$, which uses boosting algorithm [95–97].

Suppose, $(X_1, Y_1), \ldots .(X_n, Y_n)$ be the random variable in the dataset. Commonly Y_i is considered for one-dimensional death status as the response variable and X_i from multi-dimensional predictor variable taken as biomarkers HER2, ER, PR, and ERBB2. Boosting algorithm is based on the functional gradient descent (FGD) algorithm. It can be implemented on the Cox model using the negative gradient of the partial likelihood by L_2 Boosting [98]. L_2-boosting is often used in regression, particularly with the high number of covariates or predictor variables.

In this case, the loss function is defined as

$$\rho_{L_2}(Y, f) = \frac{|Y - f|^2}{2} \tag{5.21}$$

which yields population minimizer

$$f^*_{L_2}(X) = E[Y|X = x] = p(x) = P[Y = 1|X = x] \tag{5.22}$$

$f^*()$ can be estimated by the empirical risk function mentioned above with boosting, and its iteration follows sequential scheme as m^{th} iteration depends on previous $(m-1)^{th}$ iteration only in functional space. After performing classification on data, its efficiency can be measured by accuracy rate, sensitivity, and specificity. These measures can be described in terms of true positive rate, false-positive rate, true negative rate, and false-negative rate.

Chapter 6

High Dimensional Survival Data Analysis

6.1 Introduction

Survival analysis cannot be avoided in medical research, primarily in the oncology domain. There has been a trend to adopt genetic information in clinical practice in recent years. Genetic information is often required to promote personalized medicine. Personalized medicine promotes the best treatment for the patients. There is a link between survival outcomes and omics data. Omics data is a comprehensive approach for analyzing complete genetic or molecular profiles of humans and other organisms. We are commonly known as high-dimensional data. We will show how to leverage all omics data such as mutation, gene expression, and copy number alterations data for survival analysis. Particularly in the context of high dimensional data. We presented the TCGA cancer ATLAS and gene omnibus data sets for demonstration purposes. We always need to make select suitable treatments for patients by looking at patients genomic profiles. However, there is a gap in accepting this approach in routine practice. In this chapter, we present the high dimensional data to work with survival outcomes using R. The survival analysis is a mostly accepted outcome in oncology research. The statistical methods to analyze the survival analysis are well established by product-limit estimation. The product limit estimation assumed that the censoring mechanism is non-informative, but sometimes it may violate. The informative censoring procedure using a Bayesian framework is suitable to overcome this violation. The progression of disease in oncology is captured by survival analysis. Exploring the relation between gene expression and disease outcome will be obtained by conventional

DOI: 10.1201/9781003185598-6

survival analysis using R. The purpose of writing this chapter is to teach how to apply traditional survival to gene expression data analysis.

In recent years, the well-developed survival analysis and high dimensional data are now merged with increasing interest in the application of statistical modeling in clinical practice. Model selection becomes relevant to develop with the survival and high dimensional data analysis using R. So elaborate the variable selection techniques in high dimensional data.

6.2 Challenges in High Dimensional Data

Survival analysis provides a different direction from the data analysis. In simple words, it is the expected duration of time until an event occurs, including some clinical terms such as overall survival, recurrence, death status, etc. Two factors that are assumed to be most important in survival analysis are "event" i.e., whether the person is alive or dead considering death as an event, and "survival time of the patient", i.e., time of survival until any event occurs, keeping the term "censoring" in mind.

A regression model is built on these two factors to find out the relation between the survival time and one or more clinical covariates. A univariate Cox proportional model is used in this situation.

Survival analysis in high dimensional data includes data consisting of many genes within which only a small number of genes strongly indicate cancer. In contrast, other genes are not relevant in predicting the outcome. Therefore, handling such data with many irrelevant genes can be challenging and tedious. High-dimensional data generally consist of 'p' genes from 'n' subjects. Here $p \geq n$, commonly, it is observed that tens of thousands of genes are available for some hundreds of samples. These kinds of datasets are usually referred to as "microarray datasets".

Analyzing microarray data at an individual gene level usually leads to a list of many "significant" genes that can be used in the future for predicting death or a particular outcome.

The process of trying to interpret such an extensive list of genes is complicated. Thus, a variable selection from such high dimensional data has become one of the most discussed topics in statistics. Moreover, from a machine learning perspective [99], the large number [100] of genes with few samples in the datasets may cause overfitting [101] and have a negative impact on selecting significant biomarkers, which can be used for future prediction.

In the selection method of most relevant genes from a complete gene pool, we get different kinds of markers that help significantly to predict death. They are categorized as "predictive markers", "prognostic markers" and "biomarkers".

According to the National Cancer Institute, a biomarker is "a biological molecule found in blood, other body fluids, or tissues that signifies a normal or abnormal process, or of a condition or disease," such as cancer. Many selection techniques, such as penalized regression methods, can identify key predictive markers from high-dimensional data, which can be treated as a few essential biomarkers for predicting a future outcome. Penalized regression methods are a functional theoretical approach for selecting key indicators within an often substantially more extensive pool of available hands [102].

The objective of the variable selection method is to filter irrelevant genes from a large number of the gene pool. Every variable selection method has primarily two hurdles. Firstly, to find the model with the lowest prediction error rates. Secondly, to estimate the true set of covariates with non-zero regression coefficients. Recently several methods of variable selection have been developed from the L_1 regularization i.e., lasso [103] and L_1 L_2 regularization i.e., elastic net [104] regression like LARS [105], L_2 boosting with squared error loss [97]. In the current study, the primary focus is on adjusting for the regularization methods' stringency in the statistical modeling of survival data in high-dimensional settings for variable selection. After applying the regularization methods, a Cox proportional hazard model is suitable to filter out the significant genes and remove all the irrelevant genes from the selected gene pool. To overcome all these limitations and hurdles, we have developed the "SurvHiDim" package, where several functions are available. It applies all the required algorithms to remove the irrelevant genes and give a few numbers of significant genes a result.

6.3 Variable Selection in High Dimensional Data

6.3.1 Lasso selection

Lasso is a shrinkage and variable selection method primarily used in linear regression. The process introduces a specified upper bound to the sum of the absolute value of parameters. This causes the regression coefficients of certain regressors to shrink toward zero. This very much eases the process of variable selection as it helps identify the most significant predictors for the response variable. Thus the process compacts the model to a simpler and more meaningful form, even reducing prediction errors to a great extent. In other words, LASSO removes vain features from the models. In the case of positively correlated high dimensional data, it generates the correlated features. It is commonly used in feature selection technique and uses the $L_1 = \sum_j |\beta_j|$ penalty term instead of $L_2 = \sum_j \beta_j^2$, which is traditionally used in another similar procedure called ridge regression. Here $beta_j$'s are regressor coefficients. It penalized the factor toward least square or as maximum likelihood estimation. It consists of tuning constant λ_L, which is multiplied with the penalty term, i.e., the sum of absolute numbers as the parameter of interest β. The tuning constant is the penalizing factor that helps to decide how many predictors would make it to the final list. The formula becomes $\lambda_L \sum_{j=1}^{p} |\beta_j|$. If $(j = 0, 1, \ldots, p)$ then the LASSO penalized function say L_{LASSO} becomes

$$L_{LASSO} = \log L - \lambda_L \sum_{j=1}^{p} |\beta_j| \tag{6.1}$$

Final results and inferences are pretty sensitive to the variation in $lambda$. Thus it must be carefully determined. It is preferred to use the LASSO as the initial feature selection technique irrespective of correlated/non-correlated features. Using LASSO is to reduce the burden of high-dimensional data and proceeds with the selected variables.

6.3.2 Elastic net

The limitations of the LASSO is overcome by the elastic net regression. The elastic net function [104] under ordinary least square setting, is defined as

$$L(\lambda_1, \lambda_2, \beta) = |y - X\beta|^2 + \lambda_2 \sum_{j=1}^{p} \beta_j^2 + \lambda_1 \sum_{j=1}^{p} |\beta_j| \tag{6.2}$$

for some non-negative constant λ_1, λ_2, where $y = (y_1, y_2, \ldots.., y_n)$ is the centred response vector of n subjects, X is the design matrix based on p

standardized variables. The elastic net estimator is thus given by

$$\hat{\beta}_E = armin_\beta[(Y - X\beta)^T(Y - X\beta) + \lambda_2 \sum_{j=1}^{p} \beta_j^2 + \lambda_1 \sum_{j=1}^{p} |\beta_j|] \tag{6.3}$$

The two simple modifications to adjust for the high dimension and low sample sizes is available [104]. Firstly, a computational efficient algorithm is obtained due to the augmentation of X and Y leading to a sparse model. Secondly, the double shrinkage problem is overcome by a scaled $\hat{\beta}_E$. The resultant elastic net estimator $\hat{\beta}_E$ can be defined as

$$\hat{\beta}_E = armin_\beta[\beta^{`}(\frac{X^{`}X + \lambda_2 I}{1 + \lambda_2})\beta - 2y^{`}X\beta + \lambda_1 \sum_{j=1}^{p} |\beta_j|] \tag{6.4}$$

The elastic net estimator for a time to event outcome in the presence of censoring is more of interest. Let $x_i = (x_{i1}, \ldots.., x_{ip})$ be the p gene expression levels at time t_i for subject $i = 1, \ldots\ldots.n$. Due to censoring, $y_i = \min(t_i, z_i)$, where z_i is the time till the first censoring event for subject i and $\theta_i = 0$ indicate censoring and $\theta_i = 1$ otherwise.

6.3.3 Cox regression

The Cox proportional hazard regression model [3] is among the most popular method of analysis for censored survival data. We are interested in estimating the relationship between survival time and gene expression levels $X_1, \ldots\ldots., X_p$ of p genes for a given sample of size 'n'. The censoring indicator is θ_i while t_i is the survival time. The vector of gene expression levels for p genes of the ith sample is denoted by $x_i = (x_{i1}, \ldots\ldots., x_{ip})^{`}$. The Cox regression model for the hazard of cancer recurrence or death time t is defined as

$$\lambda(t) = \lambda_0(t)\exp(\beta_1 X_1 + \ldots. + \beta_p X_p) \tag{6.5}$$

$$\lambda(t) = \lambda_0(t)\exp(\beta^{`}X) \tag{6.6}$$

Where $\lambda_0(t)$ is the baseline hazard function when all other covariates are zero. While $\beta = (\beta_1, \ldots., \beta_p)$ are the regression coefficient vector,

$X = (X_1, \ldots., X_p)$ is the vector of gene expression levels with the corresponding sample values of $x_i = (x_{i1}, \ldots., x_{ip})$. The partial likelihood function is given by

$$L(\beta) = \prod_{i=1}^{n} [\frac{exp(\beta^{`} x)}{\sum_{t_j \leq t} exp(\beta^{`} x)}]^{\theta_i} \tag{6.7}$$

This algorithm can be performed with regularized regression analysis using the 'glmnet' package in R. We can regress the gene matrix to determine the discriminant and non-redundant genes whose coefficients were non-zero and supposedly were biologically relevant. Based on the hypothesis regarding the association within the genes in the microarray expression data, it can be expected that the LASSO and Elastic Net to generate a subset of discriminant genes for four different survival scenario. 'qgraph,' 'psych' and 'base' packages are helpful for visualization of correlation network plot, heatmap.

6.4 Survival and High Dimensional Data

High dimensional microarray technique helps in monitoring gene expressions on a global scale. An important application of such a technique is to obtain subsets of genes associated with the occurrence of a particular kind of disease such as cancer [106]. Several studies have been performed to determine the genetic profile of survival in non-small cell cancer, including approaches toward stratification of risk, [78, 107–109] with microarray technologies.

Microarray studies usually pertain to data with high dimensionality and low sample sizes. In today's rapidly advancing statistics, the variable selection from high dimensions has become a significant obstacle in almost every domain. Literature devoted to understanding the various varying selection techniques has evolved a lot in the past decade. The analysis is further complicated if the outcome of interest is time-to-event with censoring of outcome variable present in the microarray study [110].

Collinearity among the predictor variables becomes a notorious difficulty in high dimensional model selection. In high dimensional data, the collinearity can easily be spurious [111], which might lead to the choice of the wrong model. Indeed, intuition might be inaccurate in high dimensions. Over-fitting and model misidentification is very common in collinearity. In genomic studies, it is generally observed that only a fraction of molecules govern biological outcomes. For any disease classification or prediction problem, it is believed that biologically less than hundreds, maybe only tens of genes, are responsible for the disease. Such belief has two-fold advantages, firstly in the construction of a more reliable and sparse model with less complexity. And second, it would help clinicians and biologists to understand the mechanism at the molecular level [112]. Selection of essential genes in microarray studies by naïve but popular methods [113–116] rely on two-sample tests, which can easily miss signature genes [117] due to marginal correlation ranking [111]. In contrast, in HDLSS survival data, cox proportional hazard model [3] was proposed by several authors for variable selection. For example, Cox-based methods utilizing kernel transformations [118], threshold gradient descent minimization [119], and lasso penalization [119–121] have been proposed as well as an approach based on Bayesian variable selection [122].

It is challenging to identify any genetic pathway that could help develop biomarkers for the prognosis of lung adenocarcinoma. Our novel approach is based upon a two-step filtration of genes. Firstly, we can use LASSO and elastic-net regression to determine all the variables with non-zero regression coefficients, making the model sparse. Secondly, using Cox proportional hazard regression to determine the significant genes from the reduced pool. Such a two-step approach is supposed to produce genes that display a potential biological pathway and could further predict patients' survival.

6.5 Covariance Structure in High Dimensional Data

Covariance structure is an essential component in high-dimensional data analysis. A massive collection of statistical methods are available to work with covariance structure, including principal component analysis and linear and quadratic discriminant analysis. Other statistical approaches like cluster analysis and regression analysis also require knowledge of the covariance structure, or some aspects play a significant role in high-dimensional data analysis. Understanding the high-dimensional covariance structure, inverse, and precision matrix are essential. Sometimes, it becomes crucial in different applications with gene expression arrays and risk determinations steps with covariance structures.

6.6 Variable Selection

The recent development of high-throughput biotechnology enabled researchers to create several thousand potentially interesting measurements per subject. Different data capturing techniques open the scope of the survival analysis with clinical parameters in an extra dimension. In the field of survival analysis, the measurement becomes difficult because of the alignment of the human genome. Survival becomes extremely challenging while working with human genome data. However, it enhances our knowledge about the different diseases and leads to more accurate survival prediction models. This high-dimensional survival analysis application pushes us to challenge several thousand models and their corresponding parameters. However, the applications with high dimensional survival data provide complex computation difficulties. It is no longer possible to work with a multivariate Cox regression model when covariates greatly exceed the number of subjects. Identifying influential covariates becomes more complicated because thousands of hypotheses have to be tested simultaneously. The challenge is to control the false discoveries and make proper adjustments for the number of tests performed and needed, along with multiple testing corrections.
It works with precise ways to have gene expression measurements and survival outcomes for patients diagnosed with some disease. Some statistical models are essential to either predict survival times for individual patients based on their gene expression level or to identify genes whose expression shifts between the patients with good and bad prognoses. Simultaneously, the performance of the statistical models needed to be measured with model performance. The variable selection performed using LASSO outcome is presented in Figure 6.1.

6.6.1 Bayesian information criterion

The Bayesian Information Criterion (BIC) [123] prepares through the maximized log-likelihood with a penalty function. It dependents on sample size n. Presented as

$$\text{BIC}_m = -2\log(sup_{\theta_m \in \Theta_m} p(y|\theta_m)) + k_m \log(n) \tag{6.8}$$

There are multiple parameters involved to calculate the indexAIC AIC [124], and BIC replaces thos

$$\frac{\{\text{BIC}_{m_2}(n) - \text{BIC}_{m_1}(n)\}/2 - \log(BF_{1,2}(n))\}}{\log(BF_{1,2}(n))} \to 0 \text{ as } n \to \infty \tag{6.9}$$

The relative error of $e\{\text{BIC}_{m_2}(n) - \text{BIC}_{m_1}(n)\}/2$ is approximated $B_{1,2}$ to be $O(1)$ with some priors. The better approximation procedure is presented as [125] by to order $O(n^{-\frac{1}{2}})$, with the use of a reference prior called the unit information prior in the case of nested models. This approximation is based on the BF as it provides the Bayesian flavor and justifies it.

6.6.2 Deviance information criterion

It is difficult to count the number of parameters in a model, mainly while running with a random effect model and hierarchical specifications. These are commonly raised in the survival analysis due to working with frailty and latent class models. The existing model diagnostics criteria like AIC and BIC are burdened with several model parameters. However, the Deviance Information Criteria (DIC) reduces this model's complexity by the penalty and measuring the goodness of fit. The DIC uses the penalty and measures the goodness of fit through the posterior mean of the parameter vector. It is presented as $D(\theta) = -\log \ \text{L}(\theta; y)$ and $\hat{D} = E_{\theta|y}(D(\theta; y))$ as the posterior estimates. Now the measure of complexity is presented as $p_D = \hat{D} - D(E_{\theta|y}(\theta))$. It works by defining the posterior mean of the deviance minus the deviance of the posterior mean. Once the sample size grows, the expectation of p_D also gets equal to the number of parameters p in a fixed-effects model.

$\hat{\theta} = E_{\theta|y}(\theta)$, define

$$\text{DIC} = D(\bar{\theta}) + 2p_D = 2\hat{D} - D(\hat{\theta}) \tag{6.10}$$

It works from the posterior distribution of θ, and DIC can be easily computed with the Bayesian software package through routinely available statistical models like linear and generalized linear mixed effect models.

6.6.3 Predictive criteria

Now the BIC and DIC allow the model's information. Different model criteria focus on the ability to predict future observations. The model diagnostics criteria could broadly be differentiated into two categories as a cross-validation approach and rely on future replication. However, the objective is to provide evidence of the prediction of the statistical models.

6.7 Illustration Using R

Survival analysis on high dimensional by creating batches of variables

```
data(hnscc)
hidimsvc(7,105,5,0.05,ID=“id”,OS=“os”,
Death=“death”,PFS=“pfs”,Prog=“prog”,hnscc)
```

R output1

“Function executed properly”

“Estimate values of significant variables/DEGs on considering Death with Progression”

Variables	HR	LCL	UCL	pvalue
LRP2	1.09	1.00	1.19	0.03
C2orf72	1.07	1.00	1.14	0.04
GAL3ST1	1.13	1.03	1.24	0.00
ZMAT1	0.89	0.84	0.96	0.00
AFF3	0.88	0.82	0.95	0.00
METTL7B	1.09	1.01	1.17	0.02
PCDHA10	1.09	1.01	1.17	0.02
C1orf88	1.10	1.02	1.18	0.00
KIF1A	1.07	1.03	1.10	0.00
CTSE	0.91	0.85	0.98	0.02
CDHR5	0.84	0.72	0.97	0.02
B3GAT	1.00	0.82	0.98	0.02

R output2

"Estimate values of significant variables/DEGs considering Death without Progression data"

Variables	HR	LCL	UCL	pvalue
ZMAT1	0.89	0.84	0.96	0.00
AFF3	0.88	0.82	0.95	0.01
METTL7B	1.09	1.02	1.16	0.03
CHDH	0.95	0.89	1.01	0.03
B3GAT1	0.90	0.82	0.98	0.04

R output2

"List of variable/DEGs considering Death with Progression data" "LRP2" "C2orf72" "GAL3ST1" "ZMAT1" "AFF3" "METTL7B" "PCDHA10" "C1orf88" "KIF1A" "CTSE" "CDHR5" "B3GAT1" "List of variable/DEGs considering Death without Progression data" "ZMAT1" "AFF3" "METTL7B" "CHDH" "B3GAT1" "Estimates data for the DEGs/Variables found common between significant DEGs from data having death due to progression and data showing death without progression"

Variables	HR	LCL	UCL	pvalue
AFF3	0.88	0.82	0.95	0.01
B3GAT1	0.90	0.82	0.98	0.04
METTL7B	1.09	1.02	1.16	0.03
ZMAT1	0.89	0.84	0.96	0.00

"List of variable/ DEGs found common between significant DEGs from data having death due to progression and data showing death without progression" "AFF3" "B3GAT1" "METTL7B" "ZMAT1"

R output3

"Estimate values of significant variables/DEGs on considering Death with Progression"

Variables	HR	LCL	UCL	pvalue
LRP2	1.09	1.00	1.19	0.03
C2orf72	1.07	1.00	1.14	0.04
GAL3ST1	1.13	1.03	1.24	0.00
ZMAT1	0.89	0.84	0.96	0.00
AFF3	0.88	0.82	0.95	0.00
METTL7B	1.09	1.02	1.16	0.00
PCDHA10	1.09	1.01	1.17	0.02
C1orf88	1.10	1.02	1.18	0.00
KIF1A	1.07	1.03	1.12	0.00

"Estimate values of significant variables/DEGs considering Death without Progression data"

Variables	HR	LCL	UCL	pvalue
ZMAT1	0.89	0.84	0.96	0.00
AFF3	0.88	0.82	0.95	0.01
METTL7B	1.09	1.02	1.16	0.03
CHDH	0.89	0.89	1.01	0.03
B3GAT1	0.90	0.82	0.98	0.04

"List of variable/DEGs considering Death with Progression data" "LRP2" "C2orf72" "GAL3ST1" "ZMAT1" "AFF3" "METTL7B" "PCDHA10" "C1orf88" "KIF1A" "CTSE" "CDHR5" "B3GAT1" "List of variable/DEGs considering Death without Progression data" "ZMAT1" "AFF3" "METTL7B" "CHDH" "B3GAT1"

R output4

"Estimates data for the DEGs/Variables found common between significant DEGs from data having death due to progression and data showing death without progression"

Variables	HR	LCL	UCL	pvalue
AFF3	0.88	0.82	0.95	0.00
B3GAT1	0.90	0.82	0.98	0.02
METTL7B	1.09	1.02	1.16	0.03
ZMAT1	1.09	1.02	1.16	0.03
B3GAT1	0.89	0.84	0.96	0.00

"List of variable/ DEGs found common between significant DEGs from data having death due to progression and data showing death without progression" "AFF3" "B3GAT1" "METTL7B" "ZMAT1"

6.7.1 Data flietration with batches

Survival analysis on high dimensional by creating batches of variables

```
hidimsvc(7,105,5,0.05,ID="id",OS="os",
Death="death",PFS="pfs",Prog="prog",hnscc)
```

R output4

"Function executed properly" "Estimate values of significant variables/DEGs on considering Death with Progression"

Variables	HR	LCL	UCL	pvalue
LRP2	1.09	1.00	1.19	0.03
C2orf72	1.07	1.00	1.14	0.04
GAL3ST1	1.13	1.02	1.16	0.00
ZMAT1	0.89	0.84	0.96	0.00
AFF3	0.88	0.82	0.95	0.00
METTL7B	1.09	1.02	1.16	0.00
PCDHA10	1.09	1.01	1.17	0.02
C1orf88	1.10	1.02	1.18	0.00
AFF3	1.07	1.03	1.12	0.00
CTSE	0.91	0.85	0.98	0.02
CDHR5	0.84	0.72	0.97	0.02
B3GAT1	0.90	0.82	0.98	0.02

"Estimate values of significant variables/DEGs considering Death without Progression data"

Variables	HR	LCL	UCL	pvalue
ZMAT1	0.89	0.84	0.96	0.00
AFF3	0.88	0.82	0.95	0.01
METTL7B	1.09	1.02	1.16	0.03
CHDH	0.95	0.89	1.01	0.03
B3GAT1	0.90	0.82	0.98	0.04

"List of variable/DEGs considering Death with Progression data" "LRP2" "C2orf72" "GAL3ST1" "ZMAT1" "AFF3" "METTL7B" "PCDHA10" "C1orf88" "KIF1A" "CTSE" "CDHR5" "B3GAT1"

R output4

"List of variable/DEGs considering Death without Progression data"
"ZMAT1" "AFF3" "METTL7B" "CHDH" "B3GAT1"
"Estimates data for the DEGs/Variables found common between significant DEGs from data having death due to progression and data showing death without progression"

Variables	HR	LCL	UCL	pvalue
AFF3	0.88	0.82	0.95	0.00
B3GAT1	0.90	0.82	0.90	0.02
METTL7B	1.09	1.02	1.16	0.00
ZMAT1	0.89	0.84	0.96	0.00

Variables	UCL.y	pvalue.y
AFF3	0.95	0.01
B3GAT1	0.98	0.04
METTL7B	1.16	0.03
ZMAT1	0.96	0.00

"List of variable/ DEGs found common between significant DEGs from data having death due to progression and data showing death without progression"- "AFF3" "B3GAT1" "METTL7B" "ZMAT1" "Estimate values of significant variables/DEGs on considering Death with Progression"

Variables	HR	LCL	UCL	pvalue
LRP2	1.09	1.00	1.19	0.03
C2orf72	1.07	1.00	1.14	0.04
GAL3ST1	1.13	1.03	1.24	0.00
ZMAT1	0.89	0.84	0.96	0.00
AFF3	0.88	0.82	0.95	0.00
METTL7B	1.09	1.02	1.16	0.00
PCDHA10	1.09	1.01	1.17	0.02
C1orf88	1.07	1.03	1.12	0.00
CTSE	0.91	0.85	0.98	0.02
CDHR5	0.84	0.72	0.97	0.02
B3GAT1	0.90	0.82	0.98	0.02

R output4

"Estimate values of significant variables/DEGs considering Death without Progression data"

Variables	HR	LCL	UCL	pvalue
ZMAT1	0.89	0.84	0.96	0.00
AFF3	0.88	0.82	0.95	0.01
METTL7B	1.09	1.09	1.16	0.03
CHDH	0.95	0.89	1.01	0.03
B3GAT1	0.90	0.82	0.98	0.04

"List of variable/DEGs considering Death with Progression data"
"LRP2" "C2orf72" "GAL3ST1" "ZMAT1" "AFF3" "METTL7B" "PCDHA10" "C1orf88" "KIF1A" "CTSE" "CDHR5" "B3GAT1"
"List of variable/DEGs considering Death without Progression data"
"ZMAT1" "AFF3" "METTL7B" "CHDH" "B3GAT1"
"Estimates data for the DEGs/Variables found common between significant DEGs from data having died due to progression and data are showing death without progression."

R output4

Variables	HR	LCL	UCL	pvalue
AFF3	0.88	0.82	0.95	0.00
B3GAT1	0.90	0.82	0.90	0.02
METTL7B	1.09	1.02	1.16	0.00
ZMAT1	0.84	0.96	0.00	

Variables	UCL.y	pvalue.y
AFF3	0.95	0.01
B3GAT1	0.98	0.04
METTL7B	1.16	0.03
ZMAT1	0.96	0.00

"List of variable/ DEGs found common between significant DEGs from data having death due to progression and data showing death without progression"
"AFF3" "B3GAT1" "METTL7B" "ZMAT1"

LASSO for high dimensional data

hidimLasso(7,105,OS="os",Death="death",hnscc)

R output4

"Number of Significant DEGs are - 16",and"Sigificant Genes are - "LRP2" "GAL3ST1" "ZMAT1" "XIST" "ADH1B" "AFF3" "METTL7B" "PCDHA10" "RSPH1" "C1orf88" "KIF1A" "SGK2" "CDHR5" "B3GAT1" "RBP4" "PGC"

Survival analysis on high dimensional data

HiDimSurv(7,105,0.05,ID="id",OS="os",
Death="death",PFS="pfs",Prog="prog",hnscc)

R output4

"List of variables found significant on OS and survival event"
"LRP2" "C2orf72" "GAL3ST1" "ZMAT1" "AFF3" "METTL7B" "PCDHA10" "C1orf88" "KIF1A" "CTSE" "CDHR5" "B3GAT1" " RBP4" "PGC" "MLXIPL"
"List of variables found significant on PFS and progression event"
"ZMAT1" "AFF3" "METTL7B" "CHDH" "B3GAT1" "RBP4" "PGC"
"Estimated values for significant variables on OS and survival event"

Variables	HR	LCL	UCL	pvalue
LRP2	1.09	1.00	1.19	0.03
C2orf72	1.07	1.00	1.14	0.04
GAL3ST1	1.13	1.03	1.24	0.00
ZMAT1	0.89	0.84	0.96	0.00
AFF3	0.88	0.82	0.95	0.00

bonferroni correction factor in survival analysis

hidimSurvbon(7,105,0.05,ID="id",OS="os",
Death="death",PFS="pfs",Prog="prog",hnscc)

R output4

"List of variables found significant on OS and survival event" character(0) "List of variables found significant on PFS and progression event" character(0) "Estimates values for significant variables on OS and survival event" Variables HR LCL UCL pvalue < 0rows > (or 0-length row.names) "Estimates values for significant variables on PFS and progression event" Variables HR LCL UCL pvalue < 0rows > (or 0-length row.names) "Estamates data for the DEGs/Variables found common between significant DEGs from data having death due to progression and data showing death without progression"

R output4

Variables HR LCL UCL pvalue
<0 rows> (or 0-length row.names) "List of variable/ DEGs found common between significant DEGs from data having death due to progression and data showing death without progression" character(0) " List of variables found significant on OS and survival event" character(0) "List of variables found significant on PFS and progression event" character(0)
"Estimates values for significant variables on OS and survival event" Variables HR LCL UCL pvalue <0 rows> (or 0-length row.names)
"Estimates values for significant variables on PFS and progression event" Variables HR LCL UCL pvalue <0 rows> (or 0-length row.names)

Two step filteration using bonferroni correction and Lasso

```
hidimSurvbonlas(6,104,0.05,ID="id",OS="os",
Death="death",PFS="pfs",Prog="prog",hnscc)
```

R output4

"Function executed properly" "List of variable taking OS and survival event" "prog" "List of variable taking PFS and progression event" character(0) "Estimates values for significant variables on OS and survival event"

Variables	HR	LCL	UCL	p-value
Prog	6.96	5.38	9.01	0

"Estimates values for significant variables on PFS and progression event"

R output4

"Estamates data for the DEGs/Variables found common between significant DEGs from data having death due to progression and data showing death without progression"

Variables	HR	LCL	UCL	pvalue

UCL.y pvalue.y

$<$ 0rows $>$ (or 0-length row.names)

"List of variable/ DEGs found common between significant by DEGs from data having death due to progression and data showing death without progression" character(0)

"Number of variables or genes must be equal to or greater than 1, since it is not, LASSO cannot be appiled. Use HiDimSurv() function to get significant genes after using Cox Proportional Hazard fucntion"

Two step filteration without bonferroni correction

HiDimSurvlas(7,105,0.05,ID="id",OS="os",
Death="death",PFS="pfs",Prog="prog",hnscc)

R output4

"Number of Significant DEGs are - 6" "Sigificant Genes after applying LASSO are - " "AFF3" "B3GAT1" "METTL7B" "PGC" "RBP4" "ZMAT1"

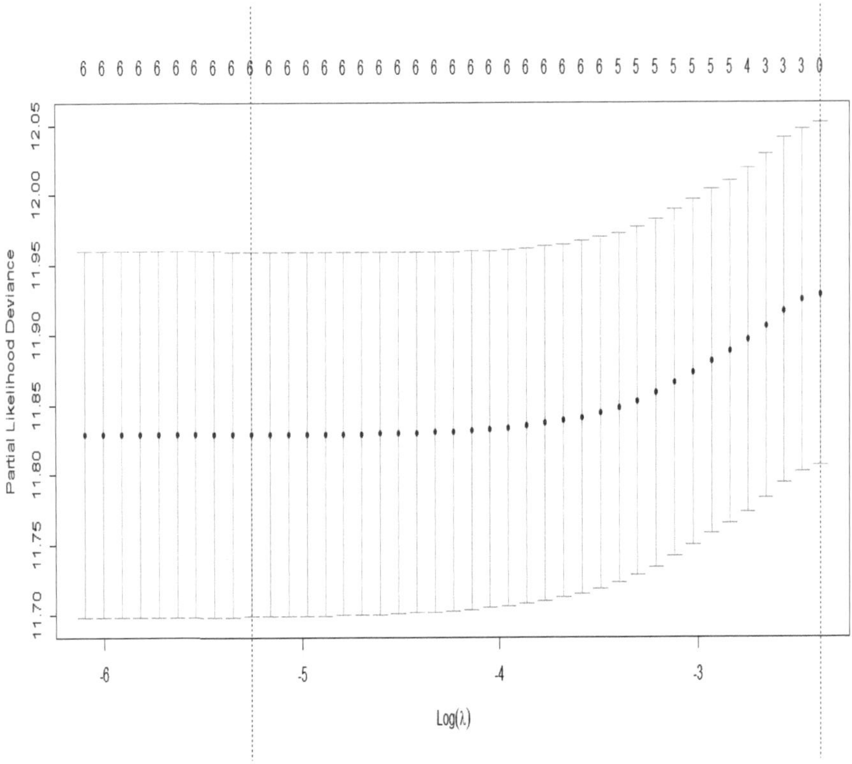

FIGURE 6.1: LASSO outcome using R.

Chapter 7

Frailty Models

7.1 Introduction

The primary application of time to event data worked with Cox proportional hazard modeling. However, the proportionality hazard assumption is considered and it is one of the limitations of the CPH. Similarly, it considered all individuals are homogenous and have a similar risk of the event. But it is not a suitable assumption at all the time. This heterogeneity is addressed by frailty modeling. It is important to consider the hazards for the events where the random effect is individual specific but it is common to all the recurrent event episodes is constant in time for any values of the covariates. However, some of these covariates get influenced by time and are popularly known as time-dependent covariates. Unless relevant covariates are measured correctly, there is always a chance of having arisen of unobserved heterogeneity.

This unobserved heterogeneity is addressed by the frailty effect [126]. The frailty is an unobserved random factor that modifies the hazard function of an individual or related individual. It is an extension of the proportionality hazard model. Commonly, the frailty models are presented with cluster failure model and recurrent event model.

A widely used function in survival analysis is the hazard function. It defines the immediate risk of the event of interest of an individual while the individual has not experienced the event previously. Now the hazard function is defined with the distribution of the time to event. Commonly, individuals determine the time to event as their own risk in terms of the probability and hazard function-younger patients having cancer live longer than older ones. Severe patients die earlier than others. All these characteristics are individual-specific. Commonly these characteristics are handled by Cox proportional

DOI: 10.1201/9781003185598-7

hazard model. The Cox model addresses even all factors; still, results are influenced by the baseline characteristics of the individuals. However, it isn't easy to work with all relevant covariates simultaneously.

The heterogeneity effect can determine the unexplained variable, and an unaccounted portion is presented as unobserved heterogeneity. So the estimated hazard for the specific covariates does not have an individual interpretation. It provides an average hazard function.

The average is determined for each time point over the individuals still alive at that time point. Now the effects of unobserved heterogeneity on lifetimes are presented as frailty in oncology research. Frailty is the unobserved individual random effect that works multiplicatively on the hazard. This estimation is spread of this unexpected effect and used to indicate the amount of the unobserved heterogeneity. Commonly, the population under study cannot be assumed as homogeneous and must be considered as heterogenous sample (mixture of individuals with different hazards).

Proportional hazards models are a class of survival models. In a Proportional hazards model, the unique effect of a unit increase in a covariate is multiplicative with respect to the hazard rate. It specifies that the ratio of the hazards for the events where the random effect is individual specific but it is common to all the recurrent event episodes is constant in time for any values of the covariates. In practical applications, some covariates effects may be time-dependent. However, if important covariates are omitted, the observed effect of modelled variables may appear time-dependent, even if, at the individual level, the effect is time constant. This phenomenon is referred to as the presence of unobserved heterogeneity.

The notion of frailty provides a convenient way of introducing unobserved heterogeneity and associations into models for survival data. Our study of interest is on a specific area in survival analysis i.e., Frailty models. In simple form, a frailty is an unobserved random proportionality factor that modifies the hazard function of an individual or of related individuals. In essence, the frailty concept goes back to work of Greenwood and Yule [127].

The term frailty itself was introduced by Vaupel et al. [126] in univariate survival models and the model was substantially promoted by its application to multivariate survival data in a seminal paper by Clayton [1978] on chronic disease incidence in families [128]. Frailty models are extensions of the proportional hazards model which is best known as the Cox model, the most popular model in survival analysis.

A Frailty model is used for univariate (independent) failure times i.e., to describe the influence of unobserved covariates in a proportional hazards model. In statistical terms, a frailty model is a random effect model for time variables, where the random effect (the frailty) has a multiplicative effect on the hazard function. One can distinguish two broad classes of frailty models:
(1) models with an univariate(independent) survival time as endpoint (e.g; Time from first heart attack to the second): It can be used to describe the influence of unobserved covariates in a proportional hazards model.
(2) models which describe multivariate(dependent) survival endpoints (e.g; competing risks, recurrence of events in the same individual, occurrence of a disease in relatives): It can be used for both survival times for individuals, (like twins or family members) and for repeated events for the same individual.

In other words, unobserved heterogeneity can be modelled explicitly via random effects or frailty models. It broadly comprises for two types of data-

1. Clustered failures (where the random effect is common to individuals from the same cluster).
2. Recurrent events (where the random effect is individual specific but it is common to all the recurrent event episodes.

There is direct relation between the probability of being alive, survival function and hazard function. It is found that some individuals differ with respect to its survival probabilities or with hazards. For e.g, Females tend to live longer than males, or in terms of hazards, the mortality rate of males is higher than that of females; patients with more severe disease tend to die earlier and have a higher hazard than less severely diseased patients.

Basically frailty models are used for modeling biomedical data, data for individuals who are correlated due to genetic relationships, etc. The purpose of using frailty models is to make sure unmeasured covariates or heterogeneity in the population. There are several methods for assessing the frailty effect. The shared frailty model considered the common random effect where individuals in a subgroup share an unobservable random covariate acting multiplicatively on the hazard rate of each group member [129].

7.2 Proportional Hazard Model

The widely used model in survival analysis is the proportional hazard model. It is commonly known as the Cox PH (Proportional hazard) model. The impact of covariates on survival outcomes is measured by the Cox PH model. Since it is a semiparametric model, so distributional assumptions are relatively low as compared to other linear types of models. The linear model performs with accelerated time model is an alternative of cox PH model with distributional assumption. Cox PH model considered the time to event and event both factor as the outcome variable, and one or more variables can be considered as covariates. The Cox PH model is defined as

$$\lambda(t|x) = \lambda_0(t)\exp(x^T\beta) \tag{7.1}$$

The term $\lambda_0(t)$ is used to define the non-parametric "baseline hazard." The covariates are presented as a vector by x is a $p \times 1$ and β is presented as a vector of the unknown regression coefficient. The frailty model is utilized as the hazard function conditional on the random effect $Z(t)$ as

$$\lambda(t|Z,x) = Z \times Y(t)\lambda_0(t)\exp(x^T\beta) \tag{7.2}$$

7.2.1 Single event frailty model

Single event frailty model refers to the scenario where individuals may experience at most one event (e.g., death). It is anticipated that the single event that, i.e., death is also presented by the frailty effect and other covariates. Covariates are possible to offer in a univariate and multivariate context. However, the single or multiple event models are suitable to handle by the marginal proportional model as an estimator and inference [130]. It corresponds to the covariate vector x_i by

$$\lambda(t|x_i) = E[Z_i|T_i \geq t, X_i]Y_i(t)\exp(x_i^T\beta)\lambda(t) \tag{7.3}$$

7.2.2 Clustered wise frailty

Individual faced the events, can be classified into different clusters. However, this cluster possibly becomes correlated as an individual nested within the groups. Clustered survival time data arise from analyses involving multiple occurrences of an event from the same Individual (e.g., repeated suicide attempts). Now, (i, j) : individual i from cluster j; Z_i : frailty shared by the individuals from cluster j. Therefore, $E[Z_i|O(t)]$ is used to define the event history of all individuals in cluster j. Denote $N_i(t)$ as the number of observed events until time t as

$$\Delta_i(t) = \sum_{i=1}^{J_j} \int_0^t Y_{ij}(s)\text{exp}(x_{ij}^T\beta)\lambda_0(s)ds \tag{7.4}$$

7.2.3 Recurrent events

In recurrent events, an individual is at risk for the same event throughout the follow-up period, despite whether an event has occurred or not. The representation of gap times is virtually identical to the clustered failures case (with individuals playing the role of a cluster and recurrent event episodes nested within individuals) and is not treated further.

$$\Delta_i(t) = \int_0^t Y_i(t)\text{exp}(x_i^T\beta)\lambda_0(s)ds \tag{7.5}$$

The frailty model is helpful if there is a heterogeneous susceptibility toward raising the risk of recurrent events.

7.3 Frailty Model

Frailty Models in Survival Analysis presents a comprehensive overview of the fundamental approaches in the area of frailty models. The concept of frailty offers a convenient way to introduce unobserved heterogeneity and associations into models for survival data. In its simplest form, frailty is an unobserved random proportionality factor that modifies the hazard function of an individual or a group of related individuals. In the univariate frailty model, the hazard of an individual with frailty Z is specified as

$$\lambda(t|Z) = Z\lambda(t) \tag{7.6}$$

The frailty Z is a random term which is assumed to have a non-negative distribution. Individuals with a higher frailty are known to be more frail and consequently expected to die earlier than other whom with the same measured covariates. For example, if the event of interest is positive outcome like recovery from certain disease then the subjects with a higher frailty are resulted to experience the positive outcome earlier than others with the same covariates. Simply conditional hazard is referred as $\Lambda(t)$.

$$\Lambda(t) = \int_0^t \lambda(s)\, ds \tag{7.7}$$

The conditional survival function for an individual with frailty Z is given by

$$S(t|Z) = \exp(-Z\Lambda(t)) \tag{7.8}$$

The survival function S(t|Z=1) corresponds to the survival of an individual with average frailty value 1. The survival of the heterogeneous population with different frailty values of individuals is called marginal survival curve which is associated with $(\Lambda(t))$. This can be obtained as $\bar{S}$(t) = E[exp(-Z($\Lambda(t)$)] $\bar{S}$(t) might be weighted average of individual survival curves if there were no covariates and the weighing depends on the distribution of Z. Therefore, marginal hazard is calculated as

$$\lambda(t) = \frac{E[Zexp(-Z\lambda(t))]}{E[exp(-Z\lambda(t))]}\lambda(t) = E[Z|T \geq t]\lambda(t) \tag{7.9}$$

This is known as weighted average of individual hazards of individuals alive at time t such that weighting depends on the distribution of Z among the individuals alive at time t.

7.3.1 Frailty distribution

There are various different distributions put forward for frailty term. In most applications it is assumed that the frailty distribution is a Gamma distribution. Consequently the selection of the frailty distribution is quite vital as the choice of the distribution of the random effect in model containing both fixed effects and random effects. Since in frailty models, the dependence between correlated observations changes over time and the frailty distribution directs how the dependence changes. Mostly frailty distributions which are studied belong to Power Variance function family and later independently studied by Hougaard [131]. The Gamma, inverse Gaussian, positive stable, and compound Poisson distribution are all members of this family.

7.3.2 Univariate frailty model

From one patient to another patient, the performance difference raised due to several causes. Many of the causes are measurable, and many of them are not. Covariates specific univariates data exploration helps to understand the direct impact of a covariate on survival outcomes. Simultaneously the presence of the univariate analysis is used to understand the frailty effect. The Cox PH model's extension by the usage of individual-level hazard function that is presented as unobservable random variable Z, as the multiplicativity on the baseline hazard function μ, presented as

$$\mu_(t, Z, X) = Z\mu_0(t)\exp(\beta^T X) \tag{7.10}$$

7.3.3 Correlated frailty model

It is extension to shared frailty model. Basically correlated frailty models are developed for the analysis of Bivariate failure time data where the two associated random variables are used to characterize the effect due to frailty for each pair. In the correlated frailty model, the frailties of individuals in a cluster are correlated but not necessarily shared. Correlated frailty models provide not only variance parameters of the frailties as in shared frailty models, but they also contain additional parameter for modeling the correlation between frailties in each group.

7.3.4 Clustered survival data

The clustered survival time function is useful for different conditions. Suppose the survival function for a specific time defined as t with the jth observations of cluster i having the time-independent covariates as Z_{ij}, frailty variate as ω_i by

$$S_{ij}(t|Z_{ij}, \omega_i) = \exp\{-\bigwedge_0(t)\omega_i e^{\beta^T Z_{ij}}\} \tag{7.11}$$

Now the term $\bigwedge$ is unspecified with cumulative baseline hazard. We can describe various options to prepare the conditional survival function.

7.3.5 Covariates

Sometimes the covariates are sampled marginally through normal distributions as the covariates are distributed with the parameter. The β covariates are obtained with distribution parameters. The covariates can be tested marginally from normal, uniform, or discrete uniform distributions by specified covariates distribution parameters. The value of β can be obtained through the genfrail as covariance parameters. Now the users can provide the relevant covariates to pass into the covariate matrix parameter. However, the covariates could come out with unlimited dimensions. The estimation process needs the number of clusters to become higher than the covariates.

7.4 Illustration

7.4.1 Diabetic retinopathy study

The application of the frailty model was found suitable in the diabetic Retinopathy Study (DRS). This study was a randomized, controlled clinical trial involving 1758 patients for seven years of follow-up. The duration of survival in this study observed with death, censoring, and dropouts. The covariates of interested were explored by statistical model. The model defined as

$$h(t_{ij}) = \alpha_i h_0(X_{ij})\exp(X_{ij}\beta) \tag{7.12}$$

with i and j being the individual-level and eye-level indicator variables, t_{ij} being the time to event for eye j of individual i, α_i being the treatment effect. X_{ij} being the treatment modality for eye j of individual i and α_i being the frailty for individual i. The flexible parametric models are modelled on the log-cumulative hazard scale

$$\log(H_{ij}) = s(\log(t_{ij}|\gamma, k_0) + X_{ij}\beta + n_i \qquad (7.13)$$

with s(.) as a spline function of log-time with parameter vector γ and knot vector k_0. The frailty distribution was modelled either a gamma or a lognormal distribution. The data set was clustered into 197 patients consisting of 394 observations.

7.4.2 Canadian health and aging study

The Clinical Frailty Scale (CFS) was introduced in the second clinical examination of the Canadian Study of Health and Aging (CSHA) as a way to summarize the overall level of fitness or frailty of an older adult after they had been evaluated by an experienced clinician [132].The CFS is a judgement-based frailty tool that evaluates specific domains including comorbidity, function, and cognition to generate a frailty score ranging from 1 (very fit) to 9 (terminally ill).

More explicitly, CFS is an ordinal hierarchical scale that numerically ranks frailty from 1 to 9, with a score of 1 being very fit, 2 well, 3 managing well, 4 vulnerable, 5 mildly frail, 6 moderately frail, 7 severely frail, 8 very severely frail, and 9 terminally ill. Precisely, the CFS is attractive as a global synthesis assessment tool that allows frailty to be defined and graded using simple clinical descriptors which are available from routine clinical assessment. National Institute for Health and Care Excellence (NICE) had witnessed that CFS helps with decision-making. The CFS is utilized to predict the outcomes of older people hospitalized with acute illnesses. The primary objective is to identify patients who are at increased risk of poor outcomes and who may not benefit from critical care interventions.

The inter-rater reliability of the CFS is very good, however, several biases may play a role in scoring, especially with clinicians who are inexperienced using the scale. This scale has largely not been validated in younger people as a disability in younger people (including both acquired, as in spinal cord injury, and life-long, as in intellectual disability) does not have the same meaning for prognosis that it does with age-related disability.

7.5 Frailty Model in Packages

Support for frailty models exists in many statistical packages such as **R, SAS, Stata** [133]. Explicitly in context to frailty mainly Gamma, Log-normal and Laplace distributions are used. Semi-parametric gamma and log-normal frailty models can be estimated in **survival** [65] package. Semi-parametric frailty models with infinitely divisible class of frailty distributions may be estimated through EM algorithm with the **frailtyEM** [134] package. Log-normal frailty models including correlated frailties can be studied with **coxme** package. Log-normal and Gamma frailty models can be estimated with the **frailtyHL** [135] package.

The pseudo-likelihood approach is implemented in the **frailtySurv** [136] package supporting distributions from Power-Variance Function (PVF) family. Parametric and flexible parametric frailty models for the Gamma and Log-normal distribution are supported by **frailtypack** [52] package which included correlated random effects, nested random effects and many others. Parametric frailty models with support for some of the PVF family distributions are implemented in the **parfm** [137] package. Stata 7 allows to explore univariate models with gamma and inverse Gaussian distributed frailty. aML 2 supports log-normal frailty models in univariate analysis. **WinBugs** is designed for analysis of shared frailty models with different frailty distribution, using Markov Chain Monte Carlo methods. In the internet, there are several SAS, GAUSS, S-Plus, and R routines available dealing with different frailty models.

Possible strategies to prevent frailty include lifestyle/behavioral factors, proper nutrition, and cognitive health maintenance. Multiple instruments exist to screen for frailty in clinical settings, and more research is needed to validate these instruments beyond their predictive value. In the univariate case they explain the deceleration of mortality rates at older ages [126] and allow for the evaluation of bias in regression coefficients when calculating the effect of observed covariates on survival in the presence of heterogeneity. This heterogeneity may be difficult to assess, but it is nevertheless of great importance. The key idea of univariate frailty models is that individuals have different frailties, and that the frailest will die earlier than the lesser frail.

Until now, frailty evaluation has been used in a variety of clinical specialties for the identification of those at highest risk for adverse outcomes and for risk stratification to assist in clinical decision making [138]. The evaluation of frailty mainly depends on identifiability conditions. For example, the characteristics of frailty identified by parametric demographic (e.g., gamma-Makeham) models differ from characteristics of frailty in biostatistical models with identifiability conditions, which in turn differ from characteristics of frailty identified from bivariate data using a correlated gamma frailty model.

The assumption that frailty is independent of observed covariates is questionable. Shared frailty models for related individuals should be used with care. Correlated frailty models maybe not be flexible enough in describing multiple correlations in clusters of larger size. To overcome current problems with the genetic analysis of durations, extended versions of susceptibility models have to be developed. For example, dynamic models of frailty need to be introduced and analyzed. Possible dependence between frailty and observed covariates should be taken into account.

7.6 Frailty and Biomarker

Continuous monitoring of the disease progression is unavoidable in cancer research. It is expected to have a continuous look at the disease progression and characteristics and monitor the disease dynamic. Disease dynamics can only be monitored by looking into important biomarkers [139]. Thus biomarker needs to be final first. Now changes in the biomarker value can be linked with a well-defined event like disease progression or relapse. Sometimes, more than one event may occur. It is also important to measure the shift of one event to another, and it is defined by multi-state modeling [140, 141]. Sometimes, different states get connected, and explore their interrelation structure is important. It is crucial to explore the dependency structure among the events to establish the impact of the biomarker as a prognostics factor. So the influence of the relevant biomarker for disease progression can be manipulated to control the disease. It is useful to prolong the survival duration and by switch the treatment plan. Suppose the biomarker is measured as continuous variable as $u_1, u_2, ..., u_n$. Now the range of the biomarker is specified as [min $u_{i,a}$] and $(a, \max\ u_i]$ by deriving the expression value into two threshold levels. Now, these two values can be generated as disease progression and treatment regimes. It lies in finding the optimum threshold value to explore the disease heterogeneity effect. Now the disease heterogeneity between the cluster of observation is often unobserved and can be modulated as a random frailty term. Recently the cancer treatment strategy is very much dependent on exploring genomic profiling. An important task is to explore survival outcomes and treatment responses for a specific biomarker to boost up the treatment strategy with some unknown heterogeneity factors. These required to explore through consideration of the frailty effect.

7.7 Illustration Using R

This "fraidm" function is available in "SurviMChd" package. It is compatible to work with time to event data. Few parameters are available in this function like m, n, Ins, Del, Time, T.min, chn, iter, data.
The description of different functions are like,
m as starting column number form where study variables to be selected.
n as ending column number till where study variables will get selected.
Ins with variable name of Institute information.
Del as Variable name containing the event information.
Time for Variable name containing the time information.
T.min with Variable name containing the time of event information.
chn is Number of MCMC chains
iter is Define number of iterations as number.
data is High dimensional data, event information given as (delta=0 if alive, delta=1 if died).
If patient is censored then $t.min$ = duration of survival. If patient is died then $t.min = 0$.
If patient is died then t = duration of survival. If patient is alive then t=NA.

High dimensional frailty data analysis

```
library(SurviMChd)
data(frailty)
fraidm(5,7,Inst="institute",Delta="del",
Time="timevar",T.min="time.min",2,7,frailty)
```

R output2

	Means(SD)	(2.5%, 97.5%)	Rhat	n.eff
b[1]	-0.38(0.20)	(-0.63,-0.12)	0.94	8
b[2]	-0.07(0.01)	(-0.09,-0.00)	14.01	2
b[3]	-0.00(0.00)	(-0.00, 0.00)	1.58	4
c[1]	78.35(2.55)	(74.88,81.68)	3.12	2
c[2]	0.03(0.03)	(-0.00,0.07)	2.82	3
omeg[1]	-0.62(0.21)	(-0.86,-0.34)	6.32	2
omeg[2]	0.69(0.57)	(-0.05,129)	7.88	2
Dev	5893.87(38.53)	(5853.00,5936.77)	13.83	2

High dimensional frailty data analysis

```
library(SurviMChd)
data(frailty)
fraidpm(5,7,Inst="institute",Delta="del",
Time="timevar", T.min="time.min",2,6,frailty)
```

Frailty effect obtained through posterior mean(SD) with Discrete Mixture Model

Parameter	Means (SD)	(2.5%, 97.5%)	Rhat	n.eff
ω_1	-5.36 (0.75)	(-6.08,-4.61)	25.38	2
ω_2	-4.07 (0.52)	(-4.58,-3.43)	6.96	2
ω_3	-5.65 (0.31)	(-5.99,-5.32)	10.98	2
ω_4	-5.35 (0.69)	(-6.06,-4.69)	20.59	2
ω_5	-4.43 (0.40)	(-4.81,-4.03)	29.50	2
ω_6	-5.88 (1.25)	(-7.06,-4.71)	75.47	2
ω_7	-5.76 (1.10)	(-6.80,-4.74)	85.81	2
ω_8	-5.36 (1.85)	(-7.07,-3.44)	22.41	2
ω_9	-4.83 (1.23)	(-5.98,-3.64)	60.53	2
ω_{10}	-5.73 (0.29)	(-6.02,-5.46)	37.71	2
ω_{11}	-5.16 (0.40)	(-5.57,-4.75)	18.24	2
ω_{12}	-5.92 (0.11)	(-6.06,-5.80)	6.60	2
ω_{13}	-5.16 (0.46)	(-5.62,-4.70)	19.88	2
ω_{14}	-6.12 (0.68)	(-6.80,-5.47)	31.55	2
ω_{15}	-6.01 (1.14)	(-7.08,-4.79)	12.32	2
ω_{16}	-4.26 (0.28)	(-4.59,-3.98)	10.85	2
ω_{17}	-4.61 (0.36)	(-4.99,-4.21)	11.81	2
ω_{18}	-4.55 (0.30)	(-4.79,-4.06)	2.95	2

The frailty effect can be obtained ω value. A total of 18 institutes are present in the "frailty" data. So the frailty effect for each institution is presented with their posterior means(SD).

High dimensional frailty data analysis with drichlet process mixture

```
library(SurviMChd)
data(frailty)
frairand(5,7,Inst="institute",Delta="del",
Time="timevar",T.min="time.min",2,6,frailty)
```

This "frairand" function is available in "SurviMChd" package. It is coupled with time to event data. Few parameters are avaliable in this function like $\mathrm{m, n, Ins, Delta, Time, T.min, chn, iter, data}$.

The description of different functions are like, m Starting column number form where study variables to be selected.n Ending column number till where study variables will get selected.Ins Variable name of Institute information.Delta Variable name containing the event information.

Time Variable name containing the time information.

T.min Variable name containing the time of event information.

chn Number of MCMC chains.

data High dimensional data, event information given as (delta=0 if alive, delta=1 if

died). If patient is censored then t.min=duration of survival. If patient is died then t.min=0. If patient is died then t=duration of survival. If patient is alive then t=NA.

Here we conidered m=5,n=7,ns="Institute",Delta="del", Time="timevar" and T.min as "time.min". number of chains defined as 2 with number of iteration as 6.

Frailty effect obtained through posterior mean(SD) with Drichlet Process Mixture

Parameter	Means(SD)	(2.5%, 97.5%)	Rhat	n.eff
c_1	76.23(1.63)	(73.72,78.22)	1.10	6
c_2	0.05(0.01)	(0.03,0.07)	1.22	6
ω_1	-0.11(0.14)	(-0.33,0.05)	2.94	2
ω_2	-0.11(0.41)	(-0.75,0.26)	0.90	6
ω_3	-0.26(0.26)	(-0.53,0.08)	0.90	6
ω_4	-0.45(0.38)	(-0.83,0.08)	1.12	6
ω_5	0.09(0.61)	(-0.92,0.66)	1.70	4
ω_6	-0.13(0.32)	(-0.46,0.37)	1.30	5
ω_7	-0.44(0.23)	(-0.72,-0.13)	1.09	6
ω_8	-0.12(0.41)	(-0.70,0.22)	2.48	3
ω_9	-0.52(0.27)	(-0.87,-0.13)	2.31	3
ω_{10}	-0.16(0.20)	(-0.45,0.06)	3.16	2
ω_{11}	-0.13(0.18)	(-0.37,0.07)	1.55	4
ω_{12}	-0.30(0.24)	(-0.59,0.08)	1.05	6
ω_{13}	-0.23(0.19)	(-0.47,-0.00)	1.50	4
ω_{14}	0.15(0.19)	(-0.02,0.46)	1.32	6
ω_{15}	-0.63(0.29)	(-0.99,-0.26)	0.81	6
ω_{16}	0.04(0.49)	(-0.60,0.58)	0.88	6
ω_{17}	0.23(0.19)	(0.01,0.53)	1.39	5
ω_{18}	0.12(0.43)	(-0.32,0.67)	2.04	3

Chapter 8

Time-Course Gene Expression Data Analysis

8.1 Introduction

Biomarker detecting tumor growth modelings are common. These are based on gene expression classification between normal and tumor cells. These findings help to identify the triggering factors for uncontrolled cell division. Similarly, tumor progressive genes can be identified only on cancer cell sequencing. Now the expression level of genes in a biological sample changed from one individual to other individuals. A changed level of expression value is circulated in cancer tissue. Later on, it can be obtained by corresponding protein expression value. Identifying those biomarkers helps implement the population-based screening tests and helps detect the disease at the initial stage [142–146].

Different approaches are available to discriminate genes between disease and control groups. The most commonly used method is the conventional t-test to present the differences of gene expression among individuals [145, 146]. Alternatively, gene expression data is segregated through Principle Component Analysis (PCA). However, it is applicable only to normal and tumor cells. The segregation between cancer patients can be obtained by Log-rank test, ANOVA, and Bayesian inference because of those deal with time-to-event outcomes. Gene expression measured with time-course observations makes the prediction impossible using conventional time-to-event data modeling and with the traditional approaches. Different research domains assumed linearity over the short interval of longitudinal data. Now the model has a random slope interval for the longitudinal data. It follows the prior distribution with successive measurements [147, 148]. Additionally, different variation with the

DOI: 10.1201/9781003185598-8

developed model incorporates the linear and quadratic behavior of the process. These models are presented as semi-parametric Bayesian versions of the local linear and quadratic state-space models. It helps us to discuss the application of the mixed effect model through state-space modeling with a longitudinal dynamic system. This work presented the Bayesian state-space models to predict the future expression levels of differently expressed genes using data collected longitudinally across samples. However, gather such information from the sample data is costly and time-consuming. In this chapter, we presented the data with microarray experiments.

The application of Bayesian state-space modeling becomes difficult while integrating with posterior distribution to estimate the parametric interest values. Now it is essential to incorporate over posterior distribution by the parametric values of interest due to the complexity of the functional form. It helps us to integrate over the unobserved process errors with posterior distribution parameters, as results in a high dimensional integrating of the equations [149]. Sometimes, it is restricted with the assumption of the linearity and normality errors, as it is possible to trace the integral value and model fitting through the application of the state-space modeling [150]. Now, the simulation with Markov Chain Monte Carlo (MCMC) helps us to retrieve approximate estimates of desired parameter values of the predictive distribution. Now the Bayesian state-space modeling becomes useful by the use of MCMC simulation from the joint posterior distribution [151]. This work is performed with BUGS software that supports the straightforward implementation of Bayesian state-space modeling using MCMC simulation. Now, the Bayesian state-space model allows the uncertainty of the unobserved system process and observed phenomenon by organizing the prior knowledge and the likelihood of experimental process [152]. However, it is robust and extensively used to explain different types of processes linked to computational statistics [153–156]. This chapter elucidates the scope of Bayesian state-space modeling on predicting the future expression values of longitudinal microarray data. We presented a robust modeling strategy to predict the future gene expression value of multiple genes. The parametric values of the predictive model and future expression levels are obtained by taking samples from the posterior joint distribution with Gibbs sampling. We can generate a 95% Credible Intervals by assuming different covariance structures like Variance Components, First-order Auto-Regressive and Unstructured variance-covariance structure to showcase the flexibility of the algorithm. The model effectively captures the increased diversity and direction of expression changes by incorporating a random intercept and subgrouping of increasing and decreasing expression values.

8.2 Microarray Data

Ever science the discovery of nucleotides and central dogma, projects are carried out to map the changes of an organism in genomic level. Micro-array studies aim to discover genes in biological samples that are differently expressed under deferent experimental or environmental conditions. A microarray may contain thousands of spots and each spot may have few million copies of identical probes or oligos that uniquely corresponds to a gene. Even though multiple platforms are available for retreating the gene sequencing, all of them share the common attribute that a sensor detects a signal from target sequences hybridized to immobilized nucleotidic probes [157].

8.2.1 Source of microarray data

Curently several thousands of microarray data is avaliable in public domain with clinical information. These provides gene expressions for different treatment comparison. Data is available at NCBI Gene Expression Omnibus (GEO) data repository under accession number GSE30531 (http://www.ncbi.nlm.nih.gov/geo). Expression values of three different samples under each treatment group was available for analysis with data of 1, 2, 4, 8, and 24 hours after treatment [158].

8.2.2 Gene expression and microarray data

Multiple methods have been evolved to analyze the gene expression data. Genomic biomarker detection is an impotent application of microarray data analysis. Multiple testing methodologies are very common in use in which they individually test each gene for evaluation association or difference in expression level with respect to independent variable. The common method in use to compare the expression level of gene in two different condition is the independent sample t-test [145, 159]. The expression values are appropriately transformed in order to keep assumptions in place. The most commonly used transformation is the logarithmic transform. This study also makes use of similar approach to select expression values which are significantly different between groups.

8.3 Model for Microarray Data

8.3.1 Bayesian state space modeling

The model development is taken place in a hierarchical way, It is assumed that the expression values say Y as an observed measurement of an underlying gene change say Z. We are trying to estimate the posterior distribution of Z where as Z is related to Y with a set of equations or models whose parameters say θ are also needed to be estimated. The observed data values are measured over a time say $t \in (1, 2, ..., T)$. The joint probability distribution of Z and θ defined as

$$P(Z, \theta|Y) \propto P(Y|Z, \theta)P(Z, \theta) \tag{8.1}$$

where the conditional distribution of Z with the knowledge of Y is given by

$$P(Z|Y) = \int P(Z, \theta|Y)d\theta \tag{8.2}$$

The integration of posterior probability equations is extremely difficult to perform analytically, hence we choose Markov Chain Monti Carlo (MCMC) method to draw samples from the posterior distribution [160].

The model building take account of the variance covariance structure of the repeated data values. Three different type of structures was under consideration, they are Variance Components (VC), first order Auto Regressive (AR(1)), and the Unstructured (US) variance covariance structure [161].

8.4 Different Covariance Structure

8.4.1 Variance Components (VC) covariance structure

The VC structure is the standard and most commonly used variance covariance structure used for the model building. Over the time $t \in (1, 2, ..., T)$, the covariance matrix of the data say Σ will be of the form,

$$\Sigma = \begin{bmatrix} {\sigma_1}^2 & 0 & \cdots & 0 \\ 0 & {\sigma_2}^2 & \cdots & 0 \\ \vdots & \vdots & \ddots & \vdots \\ 0 & 0 & \cdots & {\sigma_T}^2 \end{bmatrix}$$

$$\text{where, } 0 \leq \sigma_t$$

This model does not incorporate any type of correlation between consecutive observations which is practically a restricting assumption and not common in real life longitudinal data. Still the method is used for the joint analysis of gene expression data. Typically the assumption of VC covariance structure will simplify the estimation of variance parameters while fitting model to a longitudinal data [162].

8.4.2 First order Auto Regressive AR(1)

The AR(1) structure is useful in modeling first order temporal autocorrelations. Over the time $t \in (1, 2, ..., T)$, and assume heterogeneity in response over time, the covariance matrix of the data say Σ will be of the form,

$$\Sigma = \begin{bmatrix} {\sigma_1}^2 & \rho\sigma_1\sigma_2 & \cdots & \rho^{T-1}\sigma_1\sigma_T \\ \rho\sigma_1\sigma_2 & {\sigma_2}^2 & \cdots & \rho^{T-2}\sigma_2\sigma_T \\ \vdots & \vdots & \ddots & \vdots \\ \rho^{T-1}\sigma_1\sigma_T & \rho^{T-2}\sigma_2\sigma_T & \cdots & {\sigma_T}^2 \end{bmatrix}$$

$$\text{where, } -1 \leq \rho \leq +1 \text{ and } 0 \leq \sigma_t$$

The structure considers correlation to be highest for time adjustment data values and proportionally decreasing correlation with increase in time lag between data points. The structure assumes constant time lag between adjustment data values hence it will be inappropriate for unevenly spaced time points.

8.4.3 Unstructured (US)

The US model is the most flexible but complex type of variance covariance structure. Over the time $t \in (1, 2, ..., T)$, and assume heterogeneity in response over time, the covariance matrix of the data say Σ will be of the form,

$$\Sigma = \begin{bmatrix} {\sigma_1}^2 & \rho_{12}\sigma_1\sigma_2 & \cdots & \rho_{1T}\sigma_1\sigma_T \\ \rho_{12}\sigma_1\sigma_2 & {\sigma_2}^2 & \cdots & \rho_{2T}\sigma_2\sigma_T \\ \vdots & \vdots & \ddots & \vdots \\ \rho_{1T}\sigma_1\sigma_T & \rho_{2T}\sigma_2\sigma_T & \cdots & {\sigma_T}^2 \end{bmatrix}$$

$$\text{where,} \quad -1 \leq \rho_{..} \leq +1 \text{ and } 0 \leq \sigma_t$$

This is the most general same time most complex form of variance covariance structure. The complexity arises because it is calculating the correlation between each pair of time points available. Hence there are too many parameters need to be estimated which demands a higher sample size compared to other models.

8.5 Model Development

We assume that the probability density function of observed expression value at time t of a gene say $Y_{(t)}$ over a finite interval $t = \{1, 2, \cdots, T\}$ can be expressed as

$$\text{Log}_2(Y_{(t)}) = \{y_1, y_2, \cdots, y_T\} \sim \text{Normal}(Z_{(t)}, \Sigma)$$

where, $Z_{(t)} \in \mathbb{R}^T$ and $\Sigma \in \mathbb{R}^{T\times T}$.

The model development took place in two stages. Initially it will define a model to the data values at baseline of each expression level. Then a different model will be used to map the progression of consecutive data values. Modeling of baseline value consider that the gene expression depends on the treatment status and its interaction with time added to its normal gene expression value. The selected data of our study contains two groups such as treatment group and placebo group with upward or downward regulated genes. Let $k \in \{+1, -1, +2, -2\}$ index the regulation of group of $Log_2(.)$ transformed expression values and $\tau_{(1)}$ is the dummy value for time $t = 1$ value then the estimate of Z say $\hat{Z}$ is given by the equation,

$$\hat{Z}_{t=1} = \omega + \beta_{(k)}\tau_{(1)} \tag{8.3}$$

where ω is the random intercept or normal value for each expression and $\beta_{(k)}$ is the slope coefficient for each group. We assumed that $\omega \sim \text{Normal}(\delta, \gamma^2)$ and $\beta_{(k)} \sim \text{Normal}(\mu_{(k)}, \sigma^2_{(k)})$, where $-\infty < \delta < +\infty$, $-\infty < \mu < +\infty$, $\gamma \geq 0$, and $\sigma \geq 0$. If we have a homogeneous baseline group we can eliminate the effect of slope parameter by choosing $\tau_{(1)} = 0$.

Further expression values are adjusted in such a way that the estimated value of the current state depends on the previous state along with other variables under consideration namely the grouping and time variable. The prediction equation is of the form,

$$\hat{Z}_t = \alpha y_{(t-1)} + \beta_{(k)}(\tau_{(t)} - \alpha\tau_{(t-1)}) \tag{8.4}$$

in which α is the parameter representing the relationship of estimated present value to the previous observation. In this equation we assume that the α is a single parameter which represents the influence of previous expression value and the duration between measurements. We can choose different equations for elucidating the state and transition process depends on the additional available information's but in any case, the methodology follows the same pattern as mentioned in the article.

The set of parameters need to be estimated for the model (3) and (4) is given by $\Theta = \{\beta_{(k)}, \alpha, Z_{(t)}, \Sigma\}$ from the data which consists of values for $\{Y_{(t)}, k, \tau_{(t)}\}$.

The model adequately consider the variations between different genes, variation due to time points and variation from the up or down regulation nature of genes. It also take account for the dependence of current expression state to previous state. The above model is simple and easy to comprehend since it has a lesser number of parameters and the equation does not contain any complex functions or higher order parameters. Choosing such a model to explain the Bayesian state-space model will save significant amount of processing time during model fitting and at the same time capable of considering different source of variation for predicting future expression levels.

8.5.1 Gene selection procedure

The selection of expression values for modeling is the initial phase of study. The NCBI Gene Expression Omnibus (GEO) data repository was accessed for the data. The expression data with access number GSE30531 (http://www.ncbi.nlm.nih.gov/geo) was used for the analysis. The data was last updated on Jul 04, 2016 and it consists of repeated measurement of two groups treatment and control at 1hr, 2hr, 4hr, 8hr, and 24hr time points with three replication leads to a total of 30 arrays. Affymetrix Human Genome U133 Plus 2.0 Array (HG-U133-Plus-2) platform was used to retrieve the expression levels. Further information about the experiment and methodology can be cited from Liao et al. [158].

The data can be analyzed using R Studio for the selection of genes which has significantly different expression between groups. Due to the higher volume of data, the differences are measured separately for each time points using t-test and the codes generated from Geo2R data comparison tool [159]. Log_2 transformation is available to apply to the expressions where ever necessary. The P-value adjustment is made using Benjamini and Hochberg method [163]. The probes showing significant difference in expression level at each time points has been selected for further analysis. From each time point we select 500 expression values which are having the lowest adjusted P value. The expressions with insufficient information or significant in only less than 2 time points were removed. Using this information of the selected genes corresponding probe expression values were extracted from the GSE30531 data to create the dataset for modeling. The replication of expression values are reduced by calculating the average at each time point to simplify the algorithm for modeling.

The initial filtration of genes resulted in the selection of 85 significantly different expression values of 72 distinct genes from each group. The nature of regulation is checked by comparing the expression values at first hour and after 24 hour data. It is observed that the placebo group contains 55(64.71%) expressions with increased expression values by the end of study while the treatment group contains 33(38.82%) probes with expression values increased. 6(7.05%) of them where common among both the groups but more expressed among the treatment group. Selected expression values represent 72 distinct genes. Clinical relevance of selected expressions were not discussed in this article as the present study focused more on the designing and fitting of the model. Hence all selected expression values are considered for the modeling.

8.5.2 Model fitting and prediction

Owing to the faster advancements in technology and computation, Bayesian modeling is getting more attention now than in the past. Present study make use of Bayesian inference Using Gibbs Sampling (BUGS) software to perform the model fitting validation and prediction. The fitting of Bayesian state space model is performed using Bayesian data augmentation approach [164]. First the data is transformed into Log_2 scale and the nature of regulation ($\{-, +\}$) is determined by checking the sign of final data point – initial data point ($Sign(y_{(24)} - y_{(1)})$) for each expression values within each group. The data with no difference is consider to have +ve sign for convenience. +1 and −1 respectively represent the group of increased and decreased expression levels among placebo group and +2 and −2 respectively represent the group of increased and decreased expression values among treatment group. Samples from joint posterior distribution is drawn based on the data values through Markov Chain Monte Carlo (MCMC) simulation using OpenBUGS. Burn in iterations of 10000 was carried out to eliminate the initial value influence. Further iterations were considered for the estimation purpose. Convergence of parameters were assessed by checking the trace and density plots.

8.5.3 Parameter estimation

Parameter estimates and the credible intervals for different variance covariance structures differ slightly. The prior distribution for log transformed expression values are defined as Normal$(Z_{i,j}, \Sigma)$. Description of the resulted

TABLE 8.1: Model parameter estimates and 95% credible interval under variance components (VC) covariance structure.

Parameter	Estimate	Lower limit	Upper limit
$\beta_{(-1)}$	0.018	0.009	0.027
$\beta_{(+1)}$	0.009	0.001	0.017
$\beta_{(-2)}$	−0.005	−0.013	0.002
$\beta_{(+2)}$	0.034	0.026	0.043
α	0.962	0.940	0.982

parameters and their credible interval using Bayesian estimation over different variance covariance structures are expressed differently. The parameter estimate and credible interval under variance components covariance structure is summarized in Table 8.1. The prior distribution of $\beta_{(k)}$ is assumed to be $Normal(\mu_k, \sigma)$, and α is assumed to be Normal(0, 1). The σ is choose to be in hundreds for a wide dispersion measure for the parameter. All other parameters except from $\beta_{(-2)}$ which represent the slope of decreased expression levels among treatment group are appears to be significantly deviated from zero. The significance of α parameter represents that the estimate of current expressions are associated with the previous observed value.

The variation of expression values on 24 hr is appeared to be larger compared to other time points.

$$\hat{\Sigma}_{VC} = \begin{bmatrix} 0.005 & 0 & 0 & 0 & 0 \\ 0 & 0.003 & 0 & 0 & 0 \\ 0 & 0 & 0.002 & 0 & 0 \\ 0 & 0 & 0 & 0.078 & 0 \\ 0 & 0 & 0 & 0 & 0.787 \end{bmatrix}$$

The parameter estimate and credible interval under AR(1) covariance structure is summarized in Table 8.2. The prior distribution of the parameter remains the same as under VC structure. Under AR(1) structure the estimate of $\beta_{(+1)}$ is changed the direction when compared with VC covariance structure but still the credible interval includes 0 which shows the insignificance of the parameter. The estimate of $\beta_{(-1)}$ and $\beta_{(+2)}$ increased its vales under AR (1) covariance structure when compare with VC structure but the correlation coefficient ρ is very near to zero which shows a lesser correlation between

TABLE 8.2: Model parameter estimates and 95% credible interval under first order autoregressive (AR(1)) covariance structure.

Parameter	Estimate	Lower limit	Upper limit
$\beta_{(-1)}$	0.040	0.023	0.058
$\beta_{(+1)}$	−0.019	−0.038	0.001
$\beta_{(-2)}$	−0.066	−0.085	−0.047
$\beta_{(+2)}$	0.077	0.057	0.099
α	0.994	0.988	1.001
ρ	0.004	−1.959	1.972

consecutive time points. This may be due to the combined influence of positive and negative correlation between the consecutive expression values of different genes.

The estimated values of covariance matrix shows highest variance on 24hr data similar to VC structure. The covariance parameters show near to zero values at adjacent time points (2hr and 4hr). The estimated values shows difference if we calculate the covariance using the formula described in the methodology because estimates were selected as result of sampling. The estimate of covariance matrix under First order Auto Regressive structure is given by

$$\hat{\Sigma}_{AR(1)} = \begin{bmatrix} 0.0117 & 0.000 & 0.003 & 0.000 & 0.166 \\ 0.0000 & 0.002 & 0.000 & 0.011 & 0.000 \\ 0.0033 & 0.000 & 0.003 & 0.000 & 0.045 \\ 0.0009 & 0.011 & 0.000 & 0.095 & 0.000 \\ 0.1669 & 0.000 & 0.045 & 0.000 & 0.815 \end{bmatrix}$$

The parameter estimate and credible interval under Unstructured covariance structure is summarized in Table 8.3. The estimated parameters under UN covariance structure are very similar to those under AR(1). As mentioned above the estimate of $\beta_{(+1)}$ is insignificantly changed the direction and $\beta_{(-1)}$ and $\beta_{(+2)}$ increased its vales when compare with the estimates under VC covariance structure. α parameter estimate remain almost same throughout the estimation under different covariance.

Like other covariance matrix mentioned above the 24th hour data shows highest variability compared to other time points. The estimated covariance matrix does not have the similar elements between upper and lower diagonal. This is due to the fact that the algorithm estimates the elements separately us-

TABLE 8.3: Model parameter estimates and 95% credible interval under Unstructured (US) covariance structure.

Parameter	Estimate	Lower limit	Upper limit
$\beta_{(-1)}$	0.041	0.023	0.058
$\beta_{(+1)}$	−0.020	-0.039	0.000
$\beta_{(-2)}$	−0.065	−0.084	−0.048
$\beta_{(+2)}$	0.078	0.056	0.100
α	0.994	0.987	1.000

ing sampling. The estimate of covariance matrix under unstructured structure is given by

$$\hat{\Sigma}_{UN} = \begin{bmatrix} 0.0093 & 0.014 & 0.013 & 0.018 & 0.009 \\ 0.0086 & 0.008 & 0.011 & 0.043 & 0.013 \\ 0.0098 & 0.014 & 0.014 & 0.012 & 0.013 \\ 0.0124 & 0.134 & 0.045 & 0.015 & 0.010 \\ 0.0042 & 0.003 & 0.002 & 0.107 & 0.815 \end{bmatrix}$$

8.5.4 Prediction of gene expression

Gathering the information from the data and prior assumption the expression values are predicted for the 48 hr using the model. A sample of 50,000 is used for predicting the expressions after a burn in iteration of 1,00,000 samples. The future expression values are estimated for different variance covariance structure and the convergence is assessed by evaluating the trace plot and the probability density plot. The width of 95% Credible Intervals (CI) are compared between the estimate of each structure. It has been observed that from overall prediction of 170 expression data, about 85(50%) values showed the minimum CI length under VC covariance structure. UN covariance structure showed minimum CI length for 50(29%) values and 35(21%) predictions has minimum CI length under AR(1) covariance structure. Chi square test is used to check whether the distribution of the selected covariance structure for minimum CI are independently distributed between each group Table 8.4. The test shows a significant P value of 0.02 which is a convincing evidence that the covariance structure for the minimum CI may change between groups.

Under VC covariance structure, among control group 60%(51) of the predicted expression has the minimum credible interval length in comparison to

TABLE 8.4: Cross tabulation between study group and the covariance structure selected for minimum CI width.

	Control	Treatment	Total
VC	51 (60%)	34 (40%)	85 (50%)
AR(1)	15 (18%)	20 (24%)	35 (21%)
UN	19 (22%)	31 (36%)	50 (29%)
Total	85 (100%)	85 (100%)	170 (100%)

other variance covariance structure. Even though treatment group also has highest frequency of minimum credible interval length under VC covariance structure, the proportion is reduced to 40%(34). The possible cause for the reduction is probably due to the increase in variation of expression levels.

Under AR(1) covariance structure, it resulted minimum credible interval for 18%(15) expressions among control group and 24%(20) among treatment group. We observe a slightly increased precession for the variation of prediction among treatment group.

Under UN covariance structure, the predicted expressions are minimum for 19(22%) expression values among controls and 31(36%) among treatments under UN covariance structure. There is a nominal improvement in the precession of prediction in treatment group when compare to other structures. This may be due to the change in variability of the data values due to the treatment effect.

Microarray time course experiments typically involve gene expression measurements for thousands of genes over relatively few time points, under one (e.g., wildtype) or more biological conditions (e.g., mutant 1, mutant 2,...). The number of time points can be 3–10 for shorter and 11–20 for longer time courses. The time points at which mRNA samples are taken are usually determined by the investigator's judgement concerning the biological events of interest and are frequently irregularly spaced, although for periodic time course experiments, equally-spaced times are standard.

Measurements of mRNA abundance are based on mRNA extracted from cell lines, tissue samples or whole organisms, and in what follows we will use the general term units for the source of the mRNA. The major advantage of microarray time course studies is that they give us the ability to monitor the temporal behavior of a biological process of interest through the measurement of expression levels of thousands of genes simultaneously. This can be a powerful experimental design for identifying patterns of gene expression in the units of interest.

Time course experiments can be classified into two main categories which we term periodic and developmental. Periodic time courses include natural biological processes whose temporal profiles follow regular patterns.

In the literature, periodic time course experiments are frequently unreplicated, that is, they arise as a single series of microarray experiments, experimenters perhaps preferring to use scarce resources obtaining a finer temporal resolution, rather than repeating measurements at times already observed. In what we term developmental time course experiments, gene expression levels are measured at successive times during a developing process, for example, during the natural growth and development of, or following a treatment applied to the units. In such cases, there are usually few prior expectations concerning the form of the temporal profiles.

Following standard practice in statistics, We can further categorize time course experiments into longitudinal and cross-sectional [165]. Longitudinal experiments are those in which the mRNA samples at different times are extracted from the same unit, be it cell line, tissue or individual. This allows joining of ordinate values corresponding to observations on the same unit at different times, either by straight lines or fitted curves, to give the unit's time course for each gene, which will also be called the temporal pattern or profile. By contrast, cross-sectional time course experiments are those in which the mRNA samples at different times are extracted from different sources (units). With cross-sectional data, the individual data points can also be joined across time, using averages when there are replicate measurements, but the interpretation of the resulting curve is different. It will not correspond to any particular unit, but will be thought of as a population curve. In practice, there will be experiments exhibiting features of both longitudinal and cross-sectional types [166].

There are more cross-sectional microarray time course experiments published to date than longitudinal ones [167, 168]. This is probably because it is often infeasible to carry out longitudinal experiments because of the limited availability of mRNA from individual organisms such as laboratory mice. However, there are some exception examples about microarray data analysis with longitudinal studies [169].

In this chapter, we review methods for the design and analysis of microarray time course experiments. We can used the methods for identifying the genes of interest to the experimenter, be they genes which change over time, or genes which change differently over time between two or more biological conditions. Depending on one's perspective, this task can be viewed as a "filtering" of the genes to remove those which are not of interest, before turning to a different kind of analysis such as clustering, or it can be seen as identifying a small to moderate list of genes for validation and further characterization. We use the microarray time course data of the study to illustrate the concept of moderation and compare some of the statistics we describe below [170].

It is important to explore special characteristics of the micro-array time course data [165]. There are three principal reasons. One is the fact that microarray time series are usually so short that we cannot consider applying methods typically used to analyze time series data, e.g., ARMA, Fourier or wavelet methods, as in Diggle's and other time series books [165]. A second reason is that there are typically thousands of genes and hence thousands of short time series, all sharing the common experimental conditions. It is natural to think of analyses which have elements in common for all genes, such as the empirical Bayes (EB) methods described below. While there is some literature on EB methods in time series, we know of none involving thousands of series, as is the case here.

Finally, an important aspect of microarray gene expression data is the clustering of genes, here based on their temporal profiles. As far as we know, this problem has not arisen in the traditional time series literature, at least not in the form we meet it here. A far more relevant body of literature is that on longitudinal data analysis, for when the data are longitudinal, this is precisely the right context. When the data are not longitudinal, our context is that of many related regression models.

8.5.5 Study design

The choice of design for microarray time course experiments will depend on several factors, principally the questions the researcher wishes to address, and the available resources, including mRNA, microarrays, and related reagents. The first and most important microarray time course design question for an investigator will be whether to carry out a longitudinal or cross-sectional study [165]. While it is often possible to address the same biological question using either longitudinal or cross-sectional experiments, the major merit of longitudinal time course studies is they provide information about the temporal changes in gene expression levels within units, something that is not possible with cross-sectional studies.

8.5.6 Longitudinal over cross sectional gene expression

In statistical terms, the difference between longitudinal and cross-sectional experiments comes from the fact that gene expression measurements are typically correlated over time within units, and these correlations can be estimated and used to advantage in longitudinal studies. On the other hand, such biological correlations cannot be detected in cross-sectional time course studies, since mRNA is extracted from different sources at different times. It follows from what has just been said that if temporal changes in gene expression over time are of primary interest to the experimenter, an effort should be made to carry out a longitudinal study wherever possible.

We appreciate that in many, perhaps most cases, this may be infeasible, because of the impossibility of repeatedly sampling the mRNA from the same units. Nevertheless, a good approximation to a longitudinal study can be often realized by creating parallel, identically treated units, and sampling from different ones at different times. The difference between this kind of design and true longitudinal design depends on just how similar are the parallel, identically treated units. In some cases, they can be very similar indeed, and we observe the correlations characteristic of a genuinely longitudinal design, although their origins may be different. In such cases, these designs can be more powerful than cross-sectional studies for detecting changes. Although we might give the impression that for design purposes, the distinction between longitudinal and cross-sectional studies is straightforward, this is not really the case. There are many contexts in which hybrid studies pose more challenging design problems. For example, we might run replicate time course experiments on plants grown in a growth chamber, where each replicate consists of a series of successively sampled plants grown together under controlled conditions.

The appropriate number of full replicates, and of plants at each time within replicates given fixed resources, will depend on the relative magnitude of the different components of variation. The number of time points is usually decided by the biological background and the cost of the study. More time points permit a finer analysis of temporal patterns, e.g., a more accurate determination of the time of onset or decay of a gene's expression, but in some experiments, accuracy here of this kind is not required. Questions of interest to an investigator might concern the temporal profile of genes for one biological condition, such as a desire to identify cell-cycle regulated genes. Alternatively, interest might focus on comparison between gene profiles across two or more conditions. We might want to identify those genes which change over time in a wild-type organism, and similarly those genes which change over time in a mutant organism, and identify those genes whose temporal profiles for the wild-type and mutant are different. The latter may include many genes which are unchanging in one or the other of the two biological conditions. The way in which such questions can affect the choice of design is explained [171].

8.5.7 Short time course experiment

A short time course experiment can be regarded as a single factor experiment with time as a factor. What makes it different from other single factor experiments is the additional information from the natural ordering of time course samples. This natural ordering of levels will lead to certain comparisons being of greater interest to the researcher, and others of lesser interest. For example, comparisons of each time point mRNA sample with the baseline or differences between consecutive time points are likely to be of greater interest than comparisons between widely separated times.

Interest might focus on specific aspects of the temporal patterns of gene responses, such as monotonicity, convexity, and linearity [171]. The design of time course experiments can be considerably more complicated in the two-color comparative experiments (e.g., cDNA arrays) in comparison with single-channel experiments (e.g., Affymetrix chips), although if a common reference design is used, the two cases are fairly similar. For short two-color comparative time courses, it is possible to enumerate all the possibilities to find the optimal design. However, for those with much larger number of time points like the yeast cell cycle data, this is not feasible. There is not much literature on the design of time course experiments to describe a method for designing short cDNA time course experiments. They optimized statistical efficiency and identified so-called admissible designs, and selected efficient designs based on the effects of most interest to the biologists, the number of arrays available, and other resources. Their approach was shown to give designs better than the popular common reference design and those incorporating all possible pairwise comparisons. Optimal design for microarray time course experiments is a research topic for the future.

8.5.8 Replication

Replication is an important aspect of all statistical experimental design. As described in Yang and Speed (2003) [171], there can be three types of microarray replicates: biological replicates where mRNA samples are taken from different units; technical replicates, where mRNA samples are taken from the same unit and are split and hybridized onto different arrays; and within-array replicates, where probes are spotted in replicate on the same array. These types apply equally to longitudinal and cross-sectional time course experiments. The variation between gene expression measurements taken on these three types of replicates will be different, and in general is gene-specific. Replication is a good thing, as it provides estimates of variability relative to which temporal changes and/or condition differences can be assessed, making analysis much more straight forward. Biological replicates are generally best, as they permit the conclusions from the experiment to be extrapolated to the wider population of units from which the experimental units were obtained, something which is not possible with only technical replicates. With unreplicated experiments, the inference to a wider population is not possible, and the analysis is less straightforward, being more dependent on unverifiable assumptions, as there is no estimate of pure error which can be used. We suggest at least 3 replicates at every time point. When replicates are available, it is better to use the variation between them in any analysis, rather that just average across the replicates and proceed as with a single time course experiment. Many of the methods we describe below require replicates, and are designed to be effective with the small numbers of replicates common in this context.

8.5.9 Identifying the genes of interest

The Student's t-test was used for testing the significance of differently expressed genes under the assumption of equal variance [172]. The p-value is used to decide essential genes at ($p < 0.05$). A few are selected for all stages ($IB1, IB2$, and IIA) and progression from a set of several thousand genes. The Kaplan-Meier curve is obtained nicely through the R software [173]. Finally, the survival curves with DFS can be compared.

8.5.10 ANOVA and F-statistic

The classical approach with ANOVA is an intuitive way to select the differently expressed genes in time course data along. Sometimes, the mixed effect model with ANOVA comes as an alternative of the ANOVA [165]. One can incorporate time as a factor and possibly replicate it with others, calculating the F-statistics. Occasionally, one can include the F-statistics with the interaction term. However, the widely used method is one-way ANOVA. The application of ANOVA in the longitudinal model avoids t-test and multiple testing problems. Now, the modified ANOVA and usual two-way are adopted without testing the normality. Broadly, t-test and ANOVA are used in longitudinal data analysis. However, it is advised to avoid those techniques and adopt mixed effect modeling.

8.5.11 Moderation

Gene replicates taken at the same time varied less as compared to measurements at different time points. Typically, the repeated measurements of genes are obtained by microarray experiments. However, some gene changes are relatively small over time, and small replicates vary considerably between the F-statistics due to the underestimated denominator. Perhaps, genes are changing over time. The change shifting is more challenging to address by F-statistics and gene ranking than many other genes in expression levels. Commonly, the gene expression changes over time, but it is challenging to identify. Sometimes the single outlier leads to a lower ranking for F-statistics in comparison to other statistics. The expression level of experiment B at 10 hours is much lower than those of experiments A and C. This single outlier leads to lower rankings for F-statistic compared to all the other statistics mentioned below. However, this gene is very likely to be of interest. The gene-specific variation, variance-covariance matrices, toward the common value estimated from the whole gene set reduces the false positive and false negative estimation.

8.5.12 Gene-specific moderation

Gene-specific variances or variance-covariance matrices toward a common value estimated from the whole gene set, the total number of false positives and false negatives can usually be reduced. This is what we term moderation. The idea of moderation has entered into the analysis of microarray data in different forms [174]. It tuned that the t-statistic by adding a suitable constant to the standard deviation, the continuous being estimated by a percentile of sample standard deviations, by minimizing a coefficient of variation, respectively. Their approaches are not based on any distributional theory. There is an idea of moderation in the univariate hierarchical Bayesian mixture model for two-channel comparative experiments by smoothing gene-specific residual sample variances toward a common value [175]. Formally introduced the moderated t-statistic in the univariate general linear model setting by substituting the denominator of t-statistic with a moderated denominator.

The gene-specific moderated sample variance is defined based on some nice distributional theory, and the hyperparameter estimates derived there are shown to perform better than those into multivariate settings and introduced the MB-statistic (multivariate empirical Bayes statistic) and the T^2 statistic to rank genes in the order of differential expression in the one- and two- sample cases in the longitudinal time course context [166].

8.6 Likelihood-Based Approach

The multivariate empirical Bayes method is helpful through the updation of longitudinal time course data. Now there are different replications in this context. Commonly, the gene-specific sample variance-covariance matrix requires to take into account. Otherwise, the results become irrelevant. Now the gene-specific score-based method is meaningful with multivariate empirical

Bayes methods. It is presented as

$$w(i) = [L\hat{\beta}(i)]^T[L\hat{V}_S(i)L^T + \lambda_\omega I_{r\times r}]^{-1}[L\hat{\beta}(i)] \quad (8.5)$$

Now L is presented as an $r \times p$ matrix of rank $r, \hat{\beta}$. The estimated $p \times 1$ vector is presented by unknown regression parameters $\beta, \hat{V}_S$ with the covariance matrix estimates by $\hat{\beta}$, and λw is a positive scalar. It can be estimated as λ_w by exploratory variables. Now, this approach becomes very small, while the asymptotic theory becomes inappropriate. The likelihood ratio-based approach is assumed with gene expression value ad composed with the population means and individual deviates. It can be constructed by the likelihood-ratio-based approach, assuming gene expression values are composed of population mean and individual deviates. It is constructed that the F-statistics for longitudinal and cross-sectional data in a standard way and presented a careful treatment of the multiple testing issue. It is suggested that the moderated LR and Hotelling T^2 statistics, with the smoothing of gene-specific sample variance-covariance matrix making use of the replicate variability information across the whole gene set, and then they ranked genes according to one or the other statistic. These two statistics were shown in a simulation study to perform about as well as the MB-statistic is described below.

8.7 Empirical Bayes Approach

The empirical Bayes (EB) approach provides the power to work with thousand of micro-array genes together. We can be coupled with multivariate hierarchical normal models by incorporating conjugate prior. These are based on the derivation of the posterior odds for differential expression in the one or two-sided sample problem. It becomes helpful with longitudinal data and serves the correlation across times. Simultaneously, it becomes complex by considering the covariance setup among the longitudinal dataset. If all genes have the same number of replicates, the EB approach for the null the hypothesis that the expected profile equals 0, or the paired two-sample problem with the the null hypothesis that the expected profiles are the same is a monotonic increasing function of the statistic $\hat{T}^2 = \tilde{t}'\tilde{t}$, where

$$\tilde{t} = n^{\frac{1}{2}}\tilde{S}^{-\frac{1}{2}}\bar{X} \tag{8.6}$$

Sometimes, it becomes valid with the traditional multivariate t-statistics as the denominator is replaced by a moderated covariance matrix $\tilde{S}$. This expression becomes useful as the gene-specific covariance shares the covariance information across genes:

$$\tilde{(S)} = \frac{(n-1)S + vA}{n-1+v} \tag{8.7}$$

Here S is the gene-specific sample variance-covariance matrix; $\bar{X}$ is the gene-specific sample average time course vector; n is the number of replicates; v and A are hyperparameters estimated from the whole gene set. The EM approach with T^2 statistic for the independent two-sample problem is applicable. It is observed that there the EB-statistic achieves the lowest number of false positives and false negatives and performs about as well as the moderated Hotelling T^2 statistic. One of the values of the multivariate empirical Bayes is that it provides a natural way to estimate the gene-specific moderated sample covariance matrix, while the likelihood ratio-based approach (moderated Hotelling T^2 statistic) does not.

8.8 Illustration Using R

Bayesian mixed effect model with MCMC using R

```
library(longit)
data(repdata)
Bysmixed(m=4,n=7,t="Age",group="Gender",
chains=4,n.adapt=100,repdata)
```

R output 8.1

```
mcarray: [1] 2.112453 Marginalizing over: iteration(100),chain(4)
tau mcarray: [1] 0.4866569 Marginalizing over: iteration(100),chain(4)
```

Bayesian mixed effect model to obtain DIC

```
library(longit)
data(msrep)
BysmxDIC(m=c(4,8,12),tmax=4,t="Age",
group="Gender",chains=4,iter=1000,out="DIC",data=msrep)
```

R output 8.2

```
Variable DIC
x11 401.95
x21 396.39
x31 408.53
```

Bayesian mixed effect model to obtain HPD

```
data(msrep)
BysmxHPD(m=c(4,8,12),tmax=4,t="Age",
group="Gender",chains=4,iter=1000,out="hpD",data=msrep)
```

R output 8.3

Parameter	mu.vect	sd.vect	2.5%	97.5%	Rhat	n.eff
sigma	1.46	0.25	1.01	2.03	1.00	51
tau	0.70	0.12	0.49	0.98	1.00	51
deviance	346.02	12.00	318.992	1.01	410	

For each parameter, n.eff is a crude measure of effective sample size, and Rhat is the potential scale reduction. DIC info (using the rule, pD = var(deviance)/2) $pD = 71.7$ and DIC = 417.7
DIC is an estimate of expected predictive error (lower deviance is better).

R output 8.4

```
data(mesrep)
Bysmxms(m=4,n=7,time="Age",
group="Gender",chains=4,n.adapt=100,data=msrep)
```

R output 8.5

[[1]]σ
mcarray: [1] 1.587546
Marginalizing over: iteration(100),chain(4)
τ
mcarray: 0.6470624
Marginalizing over: iteration(100),chain(4)

Bayesian multivariate regression with independent covariance matrix for high dimensional longitudinal data

```
data(repdata)
mvncovar1(m=4,n=7,time="Age",group="Gender",
chains=10,iter=100,repdata)
```

Bayesian multivariate normal regression with unstructured covariance

```
data(repdata)
mvncovar2(m=4,n=7,time="Age",
group="Gender",chains=4,iter=100,data=repdata)
```

R output 8.6

```
data(mesrep)
Bysmxms(m=4,n=7,time="Age",
group="Gender",chains=4,n.adapt=100,data=msrep)
```

R output 8.7

sigma
mcarray:
1.58
Marginalizing over: iteration(100),chain(4)
tau
mcarray:
0.64

Chapter 9

Survival Analysis and Time-course Data Analysis

9.1 Introduction

Investigators collect a sequence of expressions from the participants in a follow-up study to analyze the pattern of essential biomarkers as a response to the treatment or occurrence of an event [176–178]. Ever since the introduction of survival modeling, the most common approach to model the relationship between the covariates and time to event is the Cox proportionality hazard models [179]. The function illustrates the association between discreetly measured covariates and the hazard function. The longitudinal measurement pattern is represented by another essential set of models called the mixed models or random effect models [180]. Mixed models portray the within and between-subject variability in a single equation and enable us not only to construct subject-specific models but also to take account of the fixed factors. The joint models take the analysis into the next stage while it uses both the Cox and Mixed models to construct the relationship between longitudinally measured covariates and the duration of an event [181].

Advancements in computation and availability of adequate software helped to promote the use of joint modeling longitudinal and time to event data in the field of biology and health research [182, 183]. The primary purpose of doing longitudinal and time to event data is to analyze the relationship between the longitudinal pattern of a covariate and the duration of the event of interest. It has been explained that the methods analyzing the longitudinal outcome alone yield bias while assessing the joint model provides more consistent estimates [184, 185]. Other than clinical and public health research, there are many other domains that use the scope of this method [186]. The

DOI: 10.1201/9781003185598-9

use of joint longitudinal and survival modeling gained significant attention in diversified fields of science. As per our review, this method has rarely been used to associate a longitudinal protein biomarker expression and duration to an event. Very recent advancements in the modeling allow the simultaneous modeling of multiple time-dependent covariates to multiple-endpoint outcomes [184].

Proteins are large molecules formed by long chains of amino acids. They are responsible for the structure, function, and regulation of the tissues and organs. The protein expression data measures the abundance of different proteins in our body. We use longitudinal protein expression values to assess the changes in protein abundance over time. While analyzing protein expression data, joint modeling will improve the efficiency of statistical modeling and reduce the bias by considering the trend of the background process rather than the individual numeric expression values [187]. The method uses the covariates trajectory to model the relationship. This will help to follow the trend of covariates even if the values are not measured until the event of study interest. Also, these characteristics make it avoid the unexpected measurement errors of longitudinal data. Also, this method can be used as a tool to trace the missing data problem [188]. On the other hand, the Cox PH model will model the association between time-independent covariates and the survival time. These characteristics enable joint modeling to effectively capture the relationship of biomarkers toward the duration of an event. There are different analysis tools, and software packages are redly available to execute the analysis. Due to the lack of familiarity among the bioscience community, the scope of the modeling technique is not substantially exploited [189, 190]. Here we illustrate the use of the existing multivariate joint modeling technique to identify the relationship between the multivariate longitudinal biomarker measurements and the hazard function of time to event outcome measurement. We use this method to establish the relationship between the longitudinally measured biomarker and the duration of relapse of cancer. The work elucidates the development and applicability of this approach in simultaneously capturing and predicting the disease-free survival duration in the presence of multiple longitudinal biomarkers.

9.1.1 Cox proportional hazard model and filtration

Analysis based on the duration of survival in relation to the occurrence of death is considered a survival analysis. The application of survival analysis has been broadened in a variety of disciplines with the definition of survival duration and event history. It stands apart as a useful tool by accommodating the censoring of the event due to non-occurrence. The event of interest is often considered as death which phrases such terminology to this vast area of study. There is a vast literature available on survival analysis which has been broadened by a large number of researchers [191–194]. However, minimal literature is devoted toward the development of statistical methods for the analysis of multi-level survival data [195]. Frailty models are also introduced and described to analyze clustered survival data [50, 194]. The hierarchical linear model (HLM) was introduced as the primary method to analyze multi-level data with continuous variables. The hierarchical generalized linear model (HGLM) was introduced to accommodate discrete outcomes. However, minimal resources of multi-level analysis are dedicated to providing a detailed discussion of multi-level survival analysis with emphasis on applications using softwares [196].

Suppose that there are n independent subjects (i.e., patients). Let T_i be the time to failure for the ith subject or "true" event time, $i = (1, ..., n)$ and Z_i is a $k \times 1$ vector of covariates. The failure time is subject to non-informative right censoring C_i. Also we assume that $\Delta_i = 0$ indicates the failure time is censored and $\Delta_i = 1$ indicates the event observed. So, the failure time T_i' is considered as $\min(T_i, C_i)$. Therefore, the observed data consist of $\{T_i', \Delta_i, Z_i\}$ for $i = 1, ..., n$. T_i and C_i are independent given Z_i for each i.

The hazard function for the ith subject is defined as,

$$\lambda_i(t|Z_i) = \lim_{h \to 0} Pr(T_i \in [t, t+h)|T_i \geq t, Z_i)/h, \qquad t > 0 \qquad (9.1)$$

For each patient, the survival duration is split into several parts based on their follow-up dates. For a patient i, the durations are computed by either time between two follow-ups or time from the last follow-up to the occurrence of disease. So, the 1st duration is calculated as 1st follow-up date to 2nd follow-up date, 2nd duration is computed as 2nd follow-up date to 3rd follow-up date, and so on. Whenever an event occurs, the duration is calculated as days between immediate previous follow-up to the disease occurrence date. Here it

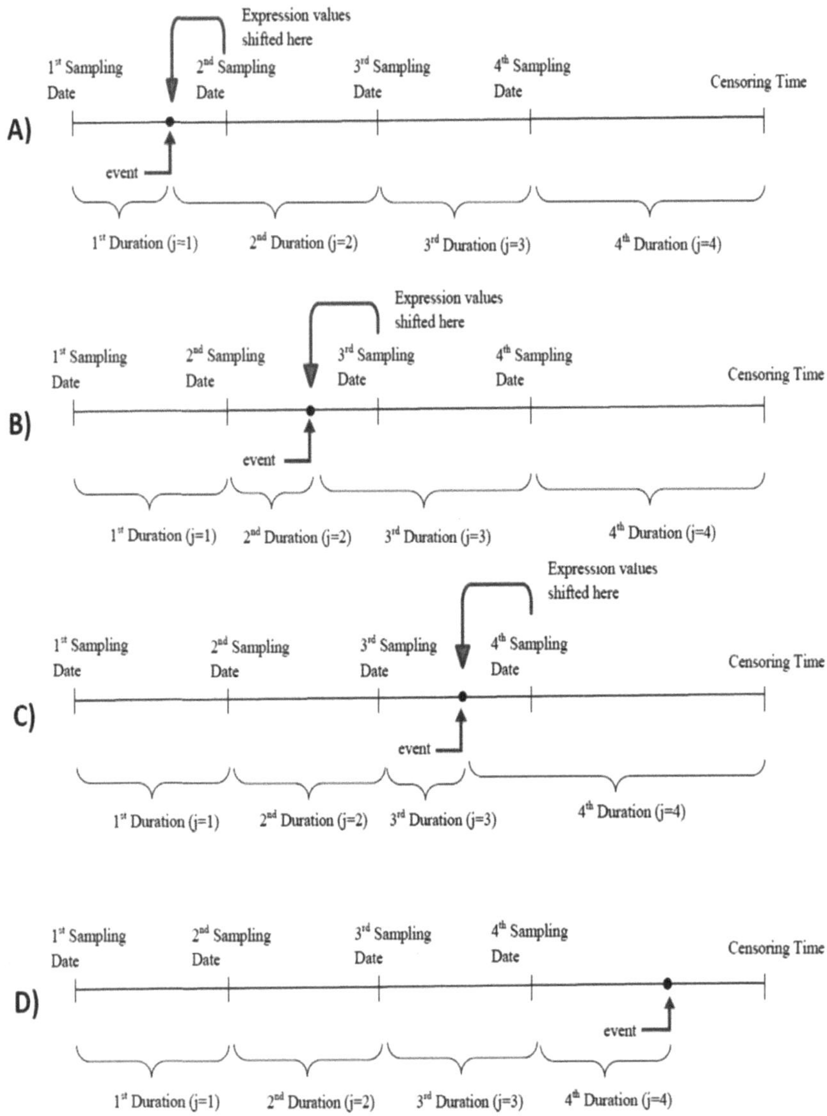

FIGURE 9.1: A-D Different scenarios of follow-up measurement and event occurrence. Events are classified as earlier occurrence of relapse/metastasis before death/censoring time. For a specific patient i, if an event occurs, then one of these A to D scenarios will happen. Otherwise patient will end his/her follow-up as censored.

is assumed that the occurrence of disease influences the gene expression. So, the value of gene expression at immediate follow-up can be considered as the value of gene expression at the event occurrence time. Hence to calculate time duration, the immediate sampling date post-disease occurrence is ignored, and the next duration is taken as the time span between 2nd immediate follow-up date and disease occurrence date. The pictorial representation of duration calculation is shown in the figure below. In this framework, the gene expression (for all the scenarios (A-E)) for each duration period j is the gene expression value measured at the beginning of the duration period, which could be the sampling date or at the time of event occurrence.

The survival duration of a patient is divided into a few parts based on the number of follow-ups. We assume that the number of follow-ups for patient i is l_i. The maximum value of l_i will be J, a pre-defined value for the study (In our study, the value of J is 4). Let, the duration between follow-ups are $t_1, t_2, ..., t_{l_i}$. So, the follow-up times are $0, t_1, t_1 + t_2, \ldots, t_1 + t_2 + ... + t_{l_i}$ The gene expressions are assigned according to the above-defined procedure. The intervals are assumed to be independent.

To analyze survival data, one of the most widely used semi-parametric method is the Cox proportional hazard model [3]. For each time point j and each gene g a proportional cox hazard model is estimated considering fixed gene expression value of gene g during the jth duration period. More precisely, for i^{th} individual, the proportional hazard model fitted at time point j for a particular gene (g) can be written in simple terms as

$$\lambda_{ij}(t|Z_{g,ij}) = \lambda_0(t)exp(Z_{g,ij}\beta_g) \tag{9.2}$$

or,

$$\log \lambda_{ij}(t|Z_{g,ij}) = \log \lambda_0(t) + Z_{g,ij}\beta_g, \quad j = 1, 2, ..., J \tag{9.3}$$

where $Z_{g,ij}$ is an indicator function defined as follows. For each gene g at time point j, the expression values are divided into deciles and we have to find the optimal threshold value of activity of these genes. For a particular p^{th} decile [p =1,2,...,9], the expression is divided into two parts, $[min\ Y_{g,j}, Y_{(g,j)p}]$ and $[Y_{(g,j)p}, max\ Y_{g,j}]$. $Y_{g,j}$ denotes the expression value for gene g at time-point j, $Y_{g,ij}$ denotes the gene expression value for i^{th} patient at j^{th} timepoint, $Y_{(g)p}$ denotes the pth decile of expression value for gene g, $Y_{(g,j)p}$ denotes the pth decile of expression value for gene g at j^{th} time point.

Thus, the indicated $Z_{(g,ij)p}(Z_{(g,ij)p}$ is the form of $Z_{g,ij}$ for p^{th} decile) is defined by,

$$Z_{(g,ij)p} = \begin{cases} 0 & \text{for } Y_{g,ij} < Y_{(g,j)p} \\ 1 & \text{for } Y_{g,ij} \geq Y_{(g,j)p} \end{cases}$$

In this way, the proportional hazard models that are fitted for different indicators on time-point j for gene g are,

$$\log \lambda_{ij}(t) = \log \lambda_0(t) + I(Y_{g,ij} > Y_{(g,j)p})\beta_{(g,j)p} \tag{9.4}$$

or,

$$\log \lambda_{ij}(t) = \log \lambda_0(t) + Z_{(g,ij)p}\beta_{(g,j)p}, \quad p = 1, 2, ..., 9 \tag{9.5}$$

where $\beta_{(g,j)p}$ is the corresponding coefficient for p^{th} decile at j^{th} time-point. For each equation, we obtain a corresponding p-value that demonstrates the significance of the covariate. The lower the p-value, the greater the significance of the covariate on response. Here, the equation with a minimum p-value gives the highest importance to the indicator, which produces the highest significant level of gene expression. Thus, for each time point, we obtain the threshold values of gene expression. Note that the proposed procedure for defining the best threshold of each gene at each time point using the minimum p-value procedure does not correct the significance level of the best threshold selected. The correction of the significance level for a series of several codings using the minimum P-value procedure has been addressed for generalized linear model [197] and Cox regression [198]. In our study, the minimum P-value procedure is used as a step to allow us to find the best cut-off for each gene at each time point in order to construct a potential biomarker whose predictive power will be evaluated using Brier score [199] (in Section 3.2).

Our interest is to classify the genes according to their effect on the disease occurrence. The gene expressions are assumed to be independent of each other. Gene expressions of some particular genes increase with the disease occurrence, while for some genes, the disease negatively affects expressions. So it is important to take into account the signs of the parameters associated with the indicator of threshold values. For the gth gene at J time-points, the parameters obtained using Cox proportional hazard model are $[\beta_{(g,1)}, \beta_{(g,2)}, ..., \beta_{(g,J)}]$. If $\beta_{(g,j)} > 0$ for all j, then the gene is defined as *Progressive Gene* and if $\beta_{(g,j)} < 0$ for all j, then the gene is defined as *Oppressive Gene*. Other than these two types of genes, there are some other genes that do not have a

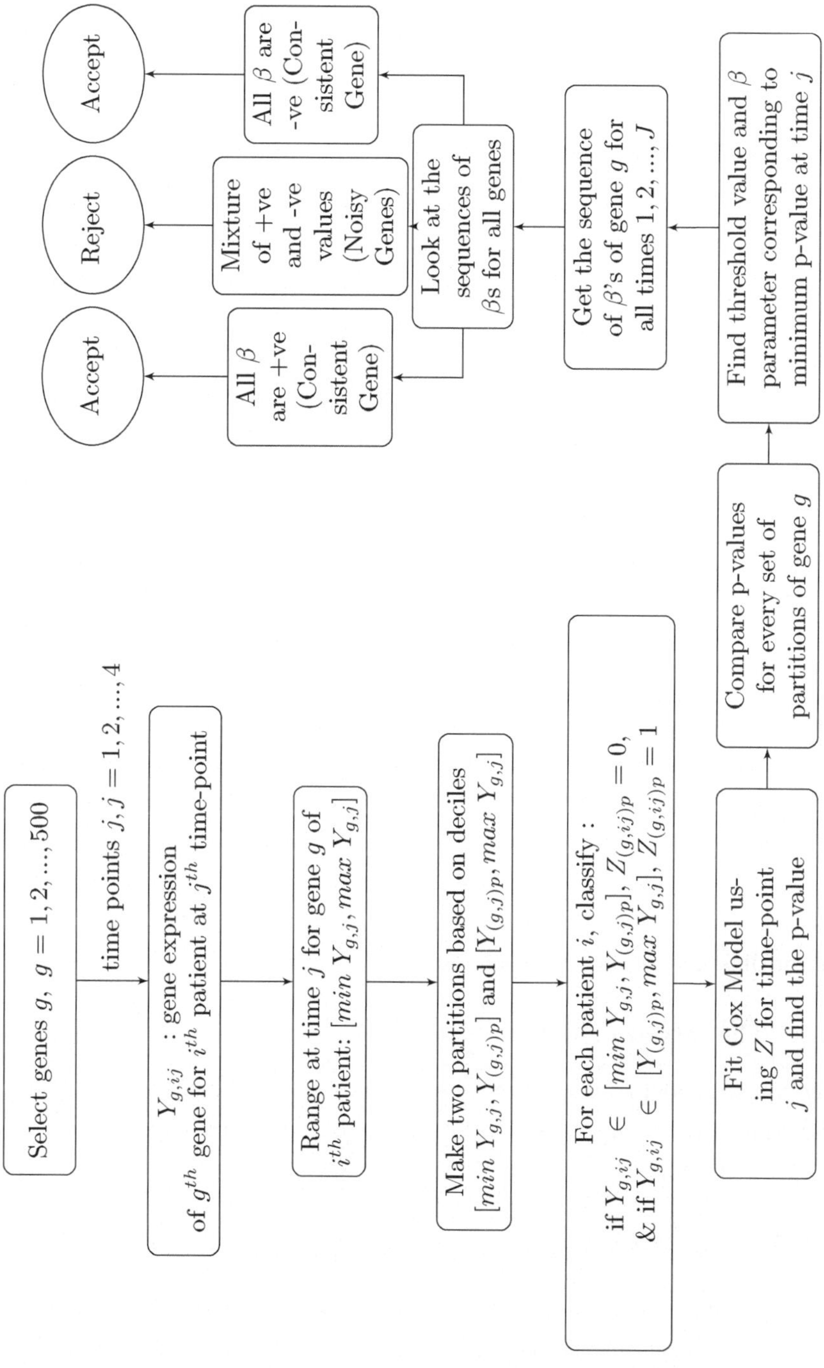

FIGURE 9.2: Flowchart to perform Classification Algorithm

consistent positive or negative association with the disease occurrence over time, i.e., $\beta_{(g,j)}$ is neither > 0 nor < 0 for all j. These genes are defined as *Volatile Genes*. The flowchart of the gene classification is shown in Figure 9.2

9.1.2 Multivariate joint model

The construction of a multivariate joint model consists of two major components. A mixed model or random-effects model captures the pattern of multivariate biomarker response, and a Cox model captures the relationship between survival and the time-fixed covariates. In joint modeling, we link these two models to construct a better survival model. The current study consider a continues set of p biomarkers for i^{th} subject, represented by $Y_i = [y_1, y_2, ..., y_p]'$ and assume that $Y \sim N(M, \Sigma)$ where M is the vector of p means and Σ is the variance-covariance matrix. Even though we assume normality, the actual model allows any mixed form of the set of biomarkers from a subject, such that each of them follows a specific distribution belonging to the exponential family. For the construction of the model, let Y_{ijk} be k^{th} observed value of the jth biomarker of ith subject $(i = 1, 2, ..., n)$, $(j = 1, 2, ..., p)$, and $k = (1, 2, 3, ..., n_{ij})$. The actual measurement durations say t_{ijk}, $t > 0$ can vary between subject to subject. Also assume that X be the set of fixed factors and Z be the set of random factors measured over time which will be used to construct the Mixed model. For the survival model, let T_i represent $min(C_i, D_i)$, where C_i is the duration to censoring and D_i is the duration to event for i^{th} subject and $t_i < T_i$. Also, e_i represents the event for i^{th} subject which takes value 1 if the event is true and 0 in case of non event and W is the set of time fixed covariates used for Cox modeling.

9.1.2.1 The mixed model

The mixed model is used to capture the subject specific pattern of longitudinally measured biomarkers. The Generalized Linear Mixed effect model is given by the formula,

$$G(Y_{ijk}) = \beta_j X_{ijk} + \gamma_{ij} Z_{ijk} + \epsilon_{ijk} \tag{9.6}$$

where $\epsilon \sim N(0, \Sigma_\epsilon)$ is the random error value, and $\gamma \sim N(0, \Sigma_\gamma)$, Σ_ϵ is the vector of the subject-specific parameters or also known as random effect parameters. Σ_γ are the covariance matrix for ϵ and γ. In this model, β_j is the vector for fixed effect parameters. $G(.)$ represents the link function used for the modeling. In our case, it is an identity function since we assume multivariate normal biomarkers. In case of violation of assumptions of the mixed effect model, there are multiple ways to trace the longitudinal outcome. Verbeke et al. compare the different methods to model the multivariate longitudinal outcome measures [200]. Generally, the purpose of this model is to build a subject-specific pattern of longitudinal biomarkers over time.

9.1.2.2 The Cox model

It is the commonly used modeling technique in survival analysis to represent the association between the hazard function and the covariates of interest. The hazard function $h(t)$ in a Cox model for i^{th} subject is represented by the equation

$$h(t) = h_0(t)\exp(\alpha W_i) \tag{9.7}$$

where $h(t)$ is the hazard function at time t, $h_0(t)$ is the baseline hazard, W_i is the vector of covariates for the subject i, and α is the vector estimated coefficients of covariates in W_i.

Even though the Cox model is generally used to model the survival, there are instances when the constant proportional hazard assumption of Cox models fails. Hickey et al. proposes the use of accelerated failure time models in such situations [201].

9.1.2.3 The Joint model

Assume the expected value of the mixed model (1) for the ith subject and j^{th} protien at any time t is represented by $\mu_{ij}(t)$ can be defined by the mixed model as,

$$\mu_{ij}(t) = \beta_j X_{ij}(t) + \gamma_{ij} Z_{ij}(t) \tag{9.8}$$

The hazard function in a joint model is defined by the equation.

$$h(t|\Xi(t), W_i) = h_0(t)\exp(\alpha W_i + \sum_{j=1}^{p} \nu\{\mu_{ij}(t)|\beta_j, \gamma_{ij}; \rho_j\} \tag{9.9}$$

where, $\Xi(t) = \{\mu_{ij}(s), 0 \leq s < t, 1 \leq j \leq p\}$ denotes the history of underlying longitudinal processes up to time t, W_i is a vector of time-fixed covariates with regression coefficients α. $\nu(.)$ specifies which components of $\mu_{ij}(t)$ are associated with the hazard at time t, and ρ_j is a vector of outcome specific regression coefficients in function $\nu(.)$ determines which components of the longitudinal outcome should include in the model [181]. This can also be called as an association parameter which determines the relationship between $\mu_{ij}(t)$ and the hazard function. The random effect parameter γ_{ij} in the model will facilitate to have a subject specific evaluation pattern.

Some notable recommendations for the definition of $\nu(.)$ by Rizopolous et al. [181] are given below,

$$\nu(.) = \rho_{(.)}\mu_{(.)}(t),$$

$\nu(.) = \rho_{1(.)}\mu_{(.)}(t) + \rho_{2(.)}\mu'_{(.)}(t)$, where $\mu'_{(.)}(t)$ is the derivative of $\mu_{(.)}$ $w.r.t$ t, $\nu(.) = \rho_{(.)} \int_0^t \mu_{(.)}(s)ds$, The joint model function considers the overall subject-specific pattern of the biomarker measurements rather than taking discrete measurement points to model the hazard function. In this study, we consider $\nu(.)$ as the simple sum function over j and k of $\rho_{(.)}\mu_{(.)}(t)$.

9.1.3 Bayesian approach in joint longitudinal and survival modeling

Likelihood approaches are computationally not effective in parameter estimation of joint models, especially when we have high dimensional latent variables. These constrain increase the scope of Bayesian parameter estimation in multivariate longitudinal and survival modeling. Complex forms of longitudinal patterns and prior information can be added while doing Bayesian parameter estimation. First proposed the estimation using Gibbs sampling by Faucett and Thomas [202]. Further developments and applications for the Bayesian parameter estimations were studied by [203–207]. Most of those studies require advanced coding to carry out the analysis in a real dataset. The JMbayes package in R, useful to work with the Bayesian approach to formulate the joint models for longitudinal and survival [208]. The estimation method follows the Markov Chain Monte Carlo (MCMC) algorithms. The recent version of the package includes the analysis of multivariate longitudinal outcomes and prediction of survival duration [183]. The theoretical development of posterior distribution is under the assumption that both the longitudinal and survival process is independent under given information about random effects. Also, the longitudinal response requires a general assumption of independence given random effects. If θ represent the set of all fixed parameters and γ represent the set of random parameters then the assumptions can be formulated in terms of probability density function $p(.)$ as,

$$p(Y_i, T_i, e_i|\gamma_i, \theta) = p(Y_i|\gamma_i, \theta)p(T_i, e_i|\gamma_i, \theta) \tag{9.10}$$

and,

$$p(Y_i|\gamma_i, \theta) = \prod_k p(Y_{ik}|\gamma_i, \theta) \tag{9.11}$$

Under this assumptions, for a given dataset say D the joint posterior distribution of γ and θ is represented as,

$$p(\gamma, \theta|D) \propto \prod_{i=1}^{n} [p(T_i, e_i|Y_i^*, W_i, \theta)p(Y_i|\gamma_i, \theta)p(\gamma_i|\Sigma_\gamma)]p(\theta|\theta_0) \tag{9.12}$$

Where Y_i^* is defined as the set of expected values of longitudinal biomarker outcomes represented by $[\mu_{i1}, \mu_{i2}, ..., \mu_{ip}]$ and θ_0 is the set of known prior distribution parameters which are used to generate the prior probabilities.

9.1.4 Description of data

For the illustration of model fitting, we use the published dataset in the Gene Expression Omnibus (GEO) database (https://www.ncbi.nlm.nih.gov/geo/) under the accession number GSE 65622. The data was published in April 2016 and last updated in August 2016. Patients received neoadjuvant chemotherapy (NACT) followed by chemoradiotherapy (CRT) and underwent treatment evaluation four weeks after CRT completion. Radical pelvic surgery was planned 2-4 weeks later. From 80 study patients, serial serum samples were collected at baseline (Sampling point = 1), following NACT post-NACT (Sampling point = 2), and at CRT completion (Sampling point = 3) post-CRT (Sampling point = 4), and at treatment evaluation (Sampling point = 5). The samples were analyzed with a high-density antibody array (AHH-BLG-1; RayBiotech Inc.) for the presence of 507 proteins. The main objective of the study was to investigate the association between circulating inflammatory factors towards the toxicity and progression-free survival of the study patients. As per the definition of the national cancer institute dictionary of cancer terms, progression-free survival is defined as *The length of time during and after the treatment of a disease, such as cancer, that a patient lives with the disease, but it does not get worse. In a clinical trial, measuring progression-free survival is one way to see how well a new treatment works. Also called PFS.* [https://www.cancer.gov/publications/dictionaries/]. The patients receive neoadjuvant chemotherapy (NCAT) followed by chemoradiotherapy (CRT) before the final pelvic surgery. Further explanation about the study setting and description of data can be cited from [176]. Figure 9.3 provides the flowchart of the treatment evaluation, and Table 9.1 provides the key points about the data applicable for our analysis.

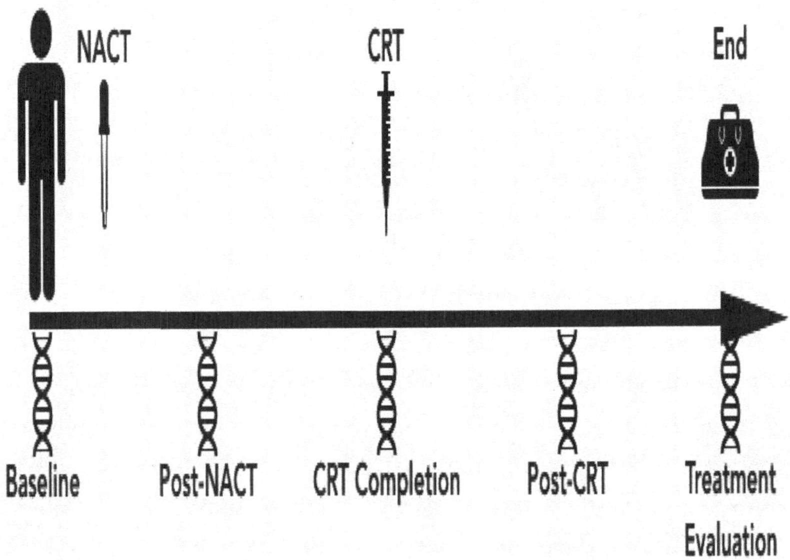

FIGURE 9.3: Treatment pattern and follow-up flowchart.

TABLE 9.1: Impotent aspects of the data use for elucidating the method.

List of baseline variables	Graded number of nearby cancerous lymph nodes (n.stage), size and extend of main tumour (t.stage), and tumour regression grading score (trg.score).
Longitudinal measurements	Expression values of proteins ADIPOQ, ANG, IGFBP2, LTBP1, LCN2, MMP9, and TNFRSF11B.
Survival measurements	Duration to censoring (C_i) , duration to progression or relapse (D_i), event status (e_i)
Missing data *[The analysis did not use any missing value imputation method for replacing missing follow-up values]*	Seventeen patients were removed from analysis due to missing baseline or no protein data Below are the number of patients with one or more missing follow-ups for each proteins Six patients have missing follow-up for ADIPOQ Five patients have missing follow-up for ANG Five patients have missing follow-up for IGFBP2 Six patients have missing follow-up for LTBP1 Sixteen patients have missing follow-up for LCN2 Two patients have missing follow-up for MMP9 Seven patients have missing follow-up for TNFRSF11B

9.2 Model Fitting

The Joint longitudinal and survival model is fitted to evaluate the impact of longitudinal biomarkers on the duration of an outcome. In the model, we test the significance of specific time-dependent and fixed parameters toward the survival time. The reference study identified significantly different 10 proteins (7 with increased fold change and 3 with decreased fold change) post NACT, 23 proteins (17 with increased fold change and 6 with decreased fold change) post CRT, and 18 proteins (15 with increased fold change and 3 with decreased fold change) in evaluation [176]. We select seven proteins (ADIPOQ, ANG, IGFBP2, TNFRSF11B, LTBP1, LCN2, MMP9) that turn out to be significantly different under all the evaluation points. The TNFRSF11B protein response was available only for 73 patients, which leads reduce the sample size from 80 to 73. The sample was further reduced to 63 patients due to missing information that was required for survival modeling. According to the final data, the ranges for indexing variables mentioned in *Section* 2 can be described as, $i = 1, 2, ..., 63$, $j = 1, 2, ..., 7$, and $k = 1, 2, ..., 5$.

Due to the absence of time-dependent covariates measurements in the data, both fixed (X_{ijk}) and random (Z_{ijk}) effects ($equation(1)$)in the mixed model fit of the protein expression values were based on the linear trend over time. Along with subject ID, the data to model the multivariate longitudinal trend contains the measurements of all seven protein expression values and the cumulative duration at which the measurements are taken. The cumulative duration is calculated as the duration from baseline measurement to the kth measurement. The multivariate linear mixed model was fitted using the "mvglmer" function in the R package "JMbayes." The package fits a multivariate mixed model under the Bayesian approach using JAGS. Point estimates of mean posterior parameters are reported in the results. To fit the Cox model, we prepare a data file with covariates and survival information. The event is defined as a relapse of disease or metastatic disease after treatment. Progression-free survival (time from baseline date to the date of local relapse or date of metastatic disease, whichever came first) was recorded for a maximum duration of follow-up for five years after surgery. Other covariates (W_i) used for fitting the survival model are T Stage (the primary tumor stage 0 means not found, 1–4 refers to the size and/or extent of the main tumor. Higher the number means larger the tumor grows), N Stage (The number and location of lymph nodes that contains cancer. Higher the number means larger the spread), TRG Score (Tumour Regression Grading score. From 0- Complete Regression to 5 Absence of regression). The model was fitted using "coxph" function in R [209] The joint model was fitted using the "mvJointModelBayes" function in the "JMbayes" package using R version 3.5.0 (2018-04-23) software. The protein expression values are assumed to follow a normal distribution. The parameter estimation was carried out using MCMC simulation, which consists of 600 burn-in iterations and 900 iterations for convergence. The chains are thinned by 300 iterations to limit the memory size. We have used a system with a 1.8GHz processor speed, and it took a total processing time of 1.42 minutes to complete. Since our purpose is to elucidate the joint longitudinal and survival model fitting, we opted for the default prior distributions and convergence test (R-hat) available in the algorithm. In real analysis, we need to perform a sensitivity analysis to evaluate the selection of prior values. The algorithm assumes the standard prior distributions in the package; that is, we use independent univariate diffuse, normal priors for all the specified association parameters and gamma prior for the precision parameters [183]. We have used the point estimates of mean posterior parameters to report the results.

9.3 Results

The results section displays the outcomes one by one in line. Since we have made many convenient modifications to the dataset and chosen the default analysis options, the final model may not be the best-fitted model for the data. All results are reported for illustrative purposes rather than drawing any clinically relevant conclusions. The figure provides the longitudinal pattern of the selected protein expression values. The proteins ADIPOQ, ANG, IGFBP2, and TNFRSF11B with increased fold change and LTBP1, LCN2, and MMP9 with decreased fold change. The figure clearly depicts the complexity of the data. As we can see, it is really hard to identify a general trend or pattern from the figure. The flowchart to perform Classification Algorithm shown in Figure 9.2. The treatment pattern and follow up flowchart given in Figure 9.3. The gene expression profile of selected proteins are detailed in Figure 9.4. The Cox proportional hazard model parameter estimates,standard error and p-values are presented with Figure Figure 9.5. Similarly, Kaplan-Meier relapse or progression free survival curves. Up Overall relapse or progression free survival, Down Stratified relapse or pro- gression free survival accoeding to Cox-model shown in Figure 9.5. Fitted survival ROC curve for assessing the predictability presented with Figure 9.5.

9.3.1 The linear mixed effect model

The fitted model considers the correlated random intercept and slope for the model. The linear mixed effect model results show the change in protein expression values longitudinally. The parameter estimates and standard error are described in Table 9.2. We have observed a significant positive trend for

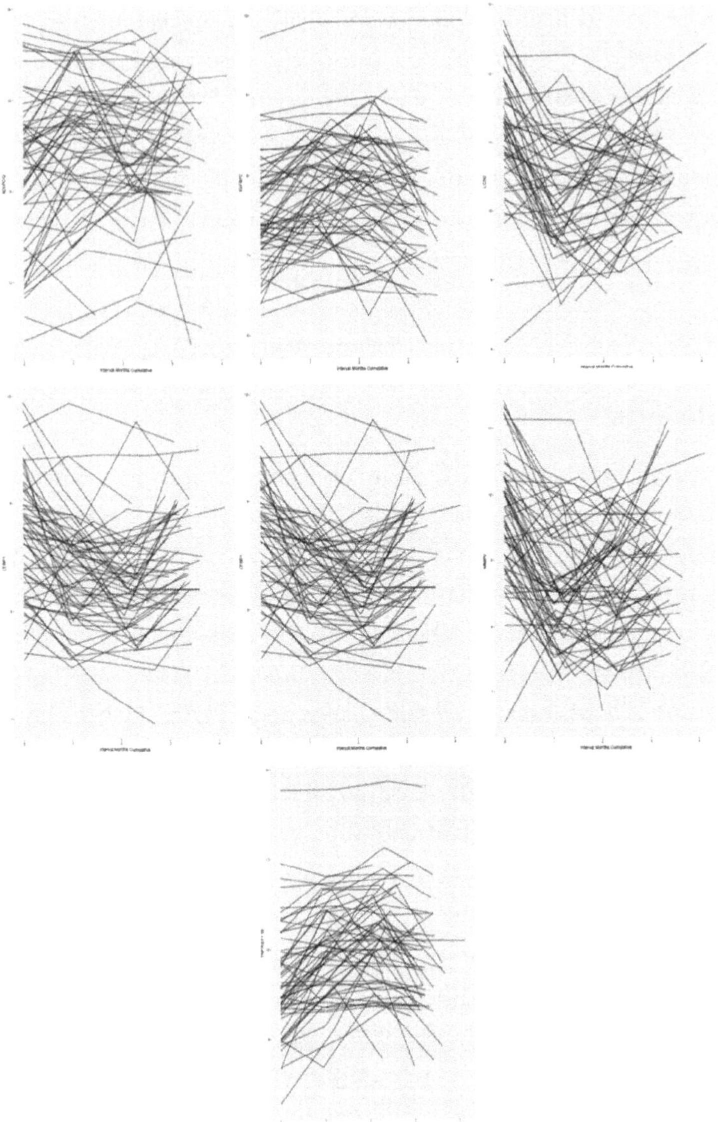

FIGURE 9.4: Expression profiling of selected proteins.

ADIPQ, ANG, and TNFRSF11B, and a non-significant positive trend was observed for IGFBP2. The proteins LTBP1, LCN2, and MMP9 show a decreasing trend over time. The table also provides the σ values of each response variable which is the estimate of the residual standard deviation of that variable [210].

TABLE 9.2: Multivariate mixed model fixed parameter estimates, standard error and p values

Protein name	Parameter	Estimate	Std err	P value
ADIPOQ				
	Intercept	13.33	0.00	0
	Cum.Month	0.06	0.00	0.00
	σ	0.39	0.00	0
ANG				
	Intercept	11.96	0.004	0
	Cum.Month	0.07	0.00	0.00
	σ	0.38	0.00	0
IGFBP2				
	Intercept	7.64	0.00	0
	Cum.Month	0.03	0.00	0.18
	σ	0.38	0.00	0
LTBP1				
	Intercept	8.56	0.00	0
	Cum.Month	−0.06	0.00	0.00
	σ	0.33	0.00	0
LCN2				
	Intercept	10.76	0.00	0
	Cum.Month	−0.09	0.00	0.02
	σ	0.69	0.00	0
MMP9				
	Intercept	9.20	0.00	0
	Cum.Month	−0.15	0.00	0.00
	σ	0.76	0.00	0
TNFRSF11B				
	Intercept	9.46	0.00	0
	Cum.Month	0.18	0.00	0
	σ	0.69	0.00	0

9.3.2 The Cox model

Cox model was fitted to model the hazard function of relapse or progression as an outcome variable with n stage, t stage, and trg score as interdependent variables. The parameter estimates and the standard error for the model are presented. Only the trg scores indicate a significant association with the event rate. The figure shows the Kaplan-Meier relapse or progression-free survival curves overall and the observed variables under the Cox model. Impotent aspects of the data use for elucidating the method are listed in Table 9.1

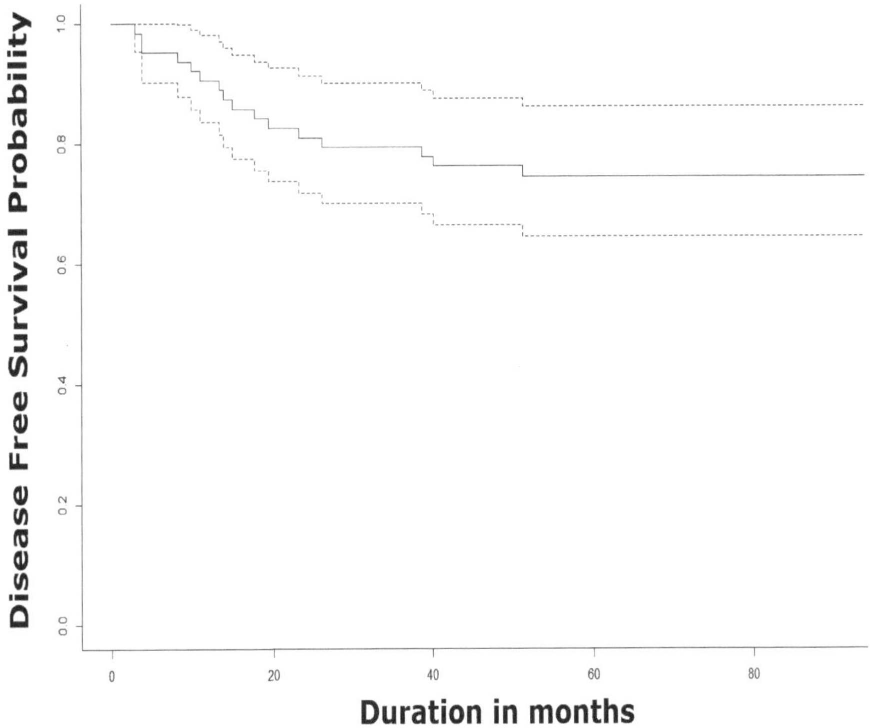

Multivariate mixed model fixed parameter estimates, standard error and p values are presented in Table 9.2. Estimates of the parameters with standard error and p values obtained through the Cox proportional hazards model are presented in Table 9.3. Estimates of the parameters with standard error and p values obtained through the joint model are presented in Table 9.4.

TABLE 9.3: Cox proportional hazards model parameter estimates, standard error and p values.

Parameter	Estimate	Std error	P value
n.stage	0.27	0.36	0.44
t.stage	−0.25	0.45	0.57
trg.score	0.70	0.25	0.00

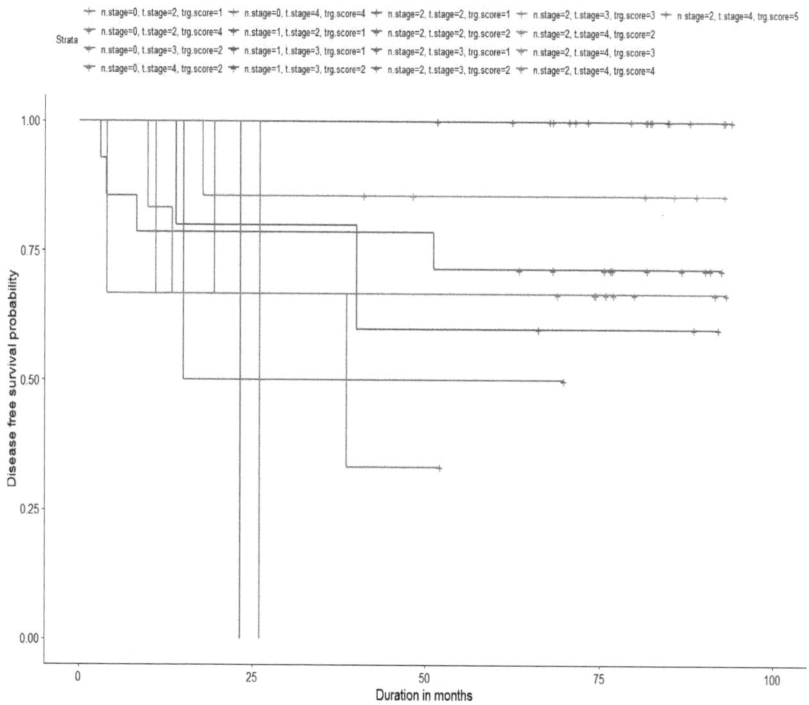

FIGURE 9.5: Kaplan-Meier relapse or progression free survival curves. *Up* Overall relapse or progression free survival, *Down* Stratified relapse or progression free survival accoeding to Cox-model

9.3.3 The joint longitudinal and survival model

The joint model output shows no significant association with any of the covariates toward the hazard function at 95% level of confidence. We observed an insignificant negative association for proteins ANG, IGFBP2, LTBP1, LCN2, and TNFRSF11B and an insignificant positive association for MMP9 and ADIPOQ. Since the parameters turn out to be insignificant, we are unable to make any valid inferences about the association. Since the objective of this chapter is only the illustration of the methodology, this drawback can be exempted.

TABLE 9.4: Joint model parameter estimates, standard error and p values.

Parameter	Estimate	Std error	P value
n.stage	0.34	0.01	0.45
t.stage	−0.44	0.01	0.41
trg.score	0.75	0.01	0.05
ADIPOQ	0.58	0.02	0.30
ANG	−0.57	0.01	0.29
IGFBP2	-0.00	0.01	0.99
LTBP1	−0.20	0.02	0.72
LCN2	−1.35	0.06	0.16
MMP9	0.86	0.04	0.20
TNFRSF11B	−0.17	0.01	0.42

9.3.4 Model validation

The model validation was carried out by predicting the relapse from 5 to 6 years. Current methods support the assessment of the models using the area under the curve (AUC) by the construction of the ROC curve. Assessing the fit showed an AUC value of 0.7107 and predicted that 43 patients were still at risk after five years. The ROC curve showed an overall good fit for predicting 5 to 6 years of survival. Now it shows the different cut-off values and corresponding misclassification measures.

The selected cut-off was at 0.9 with a sensitivity of 0.77 and specificity of 0.70. Figure 9.6 provides the plotted ROC curve for the prediction.

TABLE 9.5: The sensitivity and specificity of different cut off values for the constructed ROC cure.

Cut-off	Sensitivity	Specificity
0.75	0	0.97
0.8	0	0.97
0.85	0.12	0.86
0.87	0.59	0.81
0.90	0.76	0.69
0.95	0.92	0.16

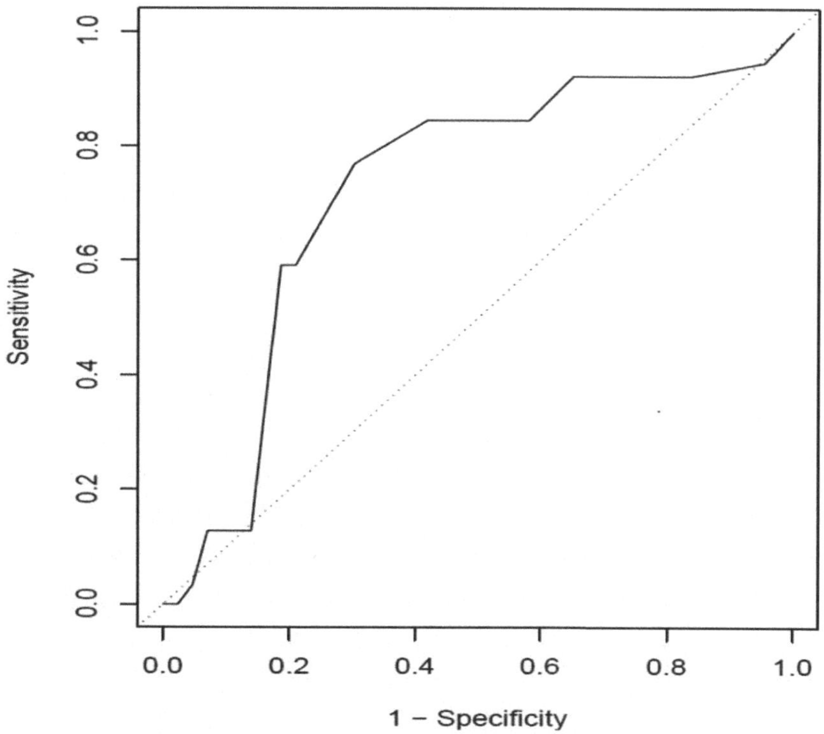

FIGURE 9.6: Fitted ROC curve for assessing the predictability.

9.4 Discussion

The increasing availability of omics data from different tools demands the use of state-of-the-art analysis techniques to come up with valid inferences. The chapter illustrates the use of multivariate joint longitudinal and survival modeling in real-life protein expression data to predict relapse-free survival. Previous works of similar modeling were primarily focused on single longitudinal biomarkers and their association with duration to events [187,190,211]. We have illustrated the use of multivariate joint longitudinal and survival modeling to analyze multiple longitudinal biomarkers simultaneously through the linear mixed effect model and the cox model setup. The sub-models (mixed model and cox model) are connected with the joint model function, which enables the researcher to concurrently take account of the trends of longitudinal outcome and duration to relapse along with other covariates. Since the model considers the overall trend of the longitudinal biomarker measurements, it can circumvent the problems of measurement bias, and missing values which are reported to be the common issues in expression data [212]. The data used in this study also contains missing follow-ups for protein expression values. For example, the patients with ID 8, 18, 19, and 24 have one or more missing follow-ups for the protein expression data. Complete data can be accessed from the Gene Expression Omnibus (GEO) database (https://www.ncbi.nlm.nih.gov/geo/) under the accession number GSE 65622. By using the Mixed effect model, the method traces the subject-specific behavior of the longitudinal biomarker over time by capturing the random effect. The inclusion of these subject-specific random parameters will enable us to estimate the subject-specific hazard for future time points, and this flexibility improves the robustness of the model.

When we analyze the result, we can see that significant influence or change in longitudinally measured covariates may not always significantly influence survival. In the mixed model result, the protein expressions of ADIPOQ, ANG, LTBP1, LCN2, MMP9, and TNFRSFT11B are influenced considerably by time, possibly as a response to the treatment. Still, the joint model result shows that none of them, along with other fixed parameters, are not significantly driving the survival duration at the 95% confidence level. At the same time, looking at the performance of the model in terms of predictability of survival by measuring the area under ROC, we obtain a considerably well-fit model. Since the objective of the chapter is only to illustrate the method of fitting the model and its validation, it is justifiable to ignore the discussion about the possible causes and impacts of significance or non-significance of specific parameters.

The recent study by Canouil M. et al. validates the feasibility of constructing the univariate joint modeling approach in genetic association studies. It identifies that joint modeling produces more precise results compared to the two-step approach [213]. Our study provides a real-life application of the same modeling technique in a multivariate longitudinal measurement situation. Joint modeling offers additional flexibility in choosing the appropriate modeling technique to capture the trend of biomarkers. Also, the association of the pattern of biomarkers toward the survival time can be represented using different functions of its longitudinal model parameters [181]. This flexibility enables us to increase the predictive accuracy of the model by trying different patterns with available data. The analysis used in this chapter assumes the normality of the longitudinal measurements. The results show that the biomarkers which indicate a significant difference in trend after treatments need not be significantly associated with the duration of the relapse but can be useful in predicting the survival probabilities. Another lately published study on the prediction of longitudinal gene expression process using Bayesian State-Space modeling also introduces a better modeling technique to trace the longitudinal process, which still requires additional computational capabilities to implement into Joint longitudinal and survival modeling [214]. The review of joint models by Hickey has already highlighted the issue of lack of software advancement has translated into reduced usage of this advanced method in medical research [201]. Joint models are well accepted statistical tool for analyzing longitudinal and survival data. This chapter elucidates the usage of multivariate joint modeling for assessing the impact of longitudinal biomarkers and the duration of an outcome. By reviewing the results, it can be observed that the fit statistics were reasonably successful in predicting the survival from 5 to 6 years. The study hints that adopting the joint model to explore the dynamics of the process will provide an areal view of the underlying process and much better estimates of survival probabilities.

Chapter 10

Features Selection in High Dimensional Time to Event Data

10.1 Introduction

Feature selection is a process of creating a subset of candidate features by dimensional reduction technique, removing irrelevant features, enhancing learning accuracy, and improving result comprehensibility [215]. An appropriate selection of the elements can guide toward the improvement of the inductive learner in terms of faster response, minimum generalization error, simplicity of the induced model, and hopefully a better understanding of the domain [216].

It is a strategy that provides us with an effective and efficient way to work with high-dimensional data along with machine-learning problems [217]. The task is to develop the procedure from gene expression data and develop a strategy by choosing more relevant genes involved in the progression of a disease. The difficulty lies in the fact that the high-dimensional datasets generally consist of a limited number of replicates, and the sample size is small [218].

Therefore, we need a criterion that uses the information to undergo a feature selection process before building any prediction model.

10.2 Different Methods in Feature Selection

Different feature selection methods usually identify the significant genes or proteins associated with the response variable. However, false-positive genes

DOI: 10.1201/9781003185598-10

are sometimes selected in the process because of a repeated number of tests and selection bias [219, 220].

A false-positive selection is an error that incorrectly determines whether a feature is accurate, and it may also happen for performing the reiteration in a process.

Selection bias is an experimental error that occurs due to inappropriate selection of sample genes that do not accurately reflect the target population [221]. It appears as a result of factors that influence subjects' participation continuously in a study. In addition, an improper feature selection procedure also leads to biases. The consequence of the biases is that the association between exposure and outcome is different for those selected in the analysis from the association among those eligible.

We can classify the feature selection methods into three categories: (I) Filter Methods, (II) Wrapper Methods, and (III) Embedded methods [222].

10.2.1 Filter method

The filter methods select features based on a performance measure and rank individual components or evaluate an entire subset of features regardless of the employed data modeling algorithm. Typical measurements for feature filtering are correlation-based feature selection, analysis of variance, chi-square, Fisher score, and multi-cluster feature selection. These measurements are generally used to assess the relationships among the response and independent variables [222, 223].

10.2.2 Wrapper method

On the contrary, the feature subsets are chosen by the performance quality of a modeling algorithm in the wrapper methods. As a black box evaluator, wrappers evaluate subsets based on the classifier performance for classification and a clustering algorithm for clustering [222,224]. Forward selection and backward elimination are two standard processes under the wrapper methods. The former process gradually incorporates the variables into larger subsets, whereas the latter one starts with all the variables taken together and progressively eliminates the least promising variables [224]. However, filter methods fail to incorporate learning algorithms, and wrapper methods utilize a learning machine to measure the quality of subsets of variables without integrating knowledge about the specific structure of the function. They can therefore be combined with any learning machine [225].

10.2.3 Embedded method

This cannot separate the learning and feature selection parts and use filter and wrapper methods to choose the independent features. Lasso, Ridge, and Elastic net regularization are some of the popular embedded methods [226, 227].

10.2.4 Other methods

Simplicity is the main advantage of filter methods, and it is not required any high computational capabilities or a long time to run the process. As opposed to filter methods, wrapper methods fit the data better by selecting relevant features for modeling rather than focusing on the association [228]. The kernel-based approach is a new way of feature selection that assists in reducing the overall dimension of the dataset using some kernel functions [229]. Another existing algorithm for feature selection is recursive feature elimination. This elimination procedure uses the list of features to train the dataset and assign weights for each of the features. The markers having the least weight are then removed using a training dataset. Sometimes, we make the assumptions of distribution and select the training parts from the actual data by random bootstrap sampling procedure using those assumptions [230].

10.2.5 Limitations of existing methods

The common drawbacks of the above methods are a false-positive or false-negative selection of features and lack of regeneration [219, 231]. Because of selection bias, sometimes the sampling data may wrongly favor or oppose a few variables. Consequently, those procedures do not appropriately select the study participants due to their presence. The results of the embedded methods also depend entirely on the taste of the training and testing datasets. Therefore, an issue of replication arises for choosing some of the variables only owing to a specific combination for selection and validation groups [232]. The re-sampling feature selection algorithm is one of the critical embedded methods that rely on sorting the important features from the entire data using re-sampling techniques. These techniques are dependent on the weight function for sorting the features. [233] defined the weight function as the proportion of each of the features in the entire iteration process. We call it as P_{weight}.

10.2.6 Re-sampling algorithm

The re-sampling algorithm is adopted to overcome the selection bias because of the taste of training and validated dataset and provide an ordered list of relevant factors according to their weights. The objective of our study is to construct a weight function that helps to select the features based on their consistency and the re-sampling method. We propose a modified version of the weight function to incorporate the concept of consistency in the feature selection method by the coefficient of variation (CV).

10.3 Distribution of Weight in Feature Selection

We can consider a time to event dataset containing n individuals and m normalized features or expression values. Also, let the list of features is denoted by $X = \{x_1, x_2, ..., x_j, ..., x_m\}$. We are interested in understanding the effect of the covariates or features on the hazard rate.

The hazard rate is the risk of suffering the event of interest given that the individual has survived up to a specific time, where the event is represented by $\{0, 1\}$ such as "0" means "no event occurred" and "1" represents "event occurred." We can use the Cox Proportional Hazards Model to illustrate the algorithm. The proportional hazards model is usually written as

$$\lambda(t \mid x_1, x_2, ..., x_m) = \lambda_0(t)\; exp(x_1\beta_1 + x_2\beta_2 + ... + x_m\beta_m) \tag{10.1}$$

$$\lambda(t \mid x_1, x_2, ..., x_m) = \lambda_0(t)\; e^{\boldsymbol{x}^T\boldsymbol{\beta}} \tag{10.2}$$

where, t is the time of an event, $\lambda_0(t)$ is known as the baseline hazard function, $\boldsymbol{\beta} = \{\beta_1, \beta_2, ..., \beta_j, ..., \beta_m\}$ is the vector of parameters and $X = \{x_1, x_2, ..., x_j, ..., x_m\}$ is a set of normalized continuous predictors.

We use the partial likelihood function for estimation of the parameters. Assuming the sample consists of n independent subjects and x_{ij} represents the value of X data matrix for the ith patient corresponding to the jth covariate or normalized predictor, $i = 1, 2, .., n$ and $j = 1, 2, ..., m$. The form of the partial likelihood is generally represented by $PL(\beta)$ and defined as,

$$PL(\beta) = \prod_{l=1}^{d} \left[\frac{exp(\boldsymbol{x}_{i(t_l)}^T \boldsymbol{\beta})}{\sum_{i=1}^{n} exp(\boldsymbol{x}_i^T \boldsymbol{\beta}) Y_i(t_l)} \right] \tag{10.3}$$

where, $t_1, t_2, ..., t_d$ are the distinct event times; $Y_i(t_l)$ denotes the indicator for whether or not the ith patient is at risk at time t_l, then $Y_i(t_l) = 1$ if the patient i is at risk at t_l, otherwise zero; $Y(t_l) = \sum_{i=1}^{n} Y_i(t_l)$ be the total number of entire sample who are at risk at time t_l; and $\boldsymbol{x}_{i(t_l)}^T$ be the vector of covariates corresponding to the patient i at time t_l.

In this illustration, we take the logarithm of the partial likelihood function. Then, the unknown parameters are estimated by maximizing the partial log-likelihood w.r.to $\boldsymbol{\beta}$.

The re-sampling feature selection algorithm begins with selecting the features analogous to the filter methods. When the number of features is superior to the full sample size, removing the unworthy parts from the data is required. This preliminary screening can be performed through any filter method. Since the illustration is treated for prediction purposes as well as the selection of relevant features using the weight function approach, one should not be concerned about the false-positive selection of variables at this time of analysis. The primary screening will assist us in eliminating the unnecessary features in the next steps, and after that, the dataset is randomly split into training and testing groups. Let the training set and the validation set be denoted by T_i and V_i, respectively. Also, let S_i is the splitting scenario, and F_i be the set of variables selected using S_i and a suitable algorithm. The process is repeated a fixed number of times (say, N) using different training and testing datasets. It would be adequate to take sufficiently large N to get better results. For each repetition, the T_i and V_i dataset are randomly chosen, and the resulting variables save to a dataset for further scrutiny.

The modified weight function for the feature x_j is defined as

$$w_j^* = exp[-\frac{1}{2}\frac{\sigma_j/\mu_j}{R}(1-\frac{n_j}{N})] \quad (10.4)$$

where, $j = 1, 2, ..., m$, and μ_j and σ_j be the respective mean and standard deviation of the jth feature.

Assuming the feature x_j is selected in the model n_j times after repeating the process N times ($n_j \leq N$), then $w_j = \frac{n_j}{N}$ is defined as the proportion of times the variable x_j selected to the model. One of the usage statistics for comparing the degree of variation from one data series to another is the coefficient of variation, even if the means are drastically different from one another. Here we are trying to incorporate the concept of consistency through the CV measure. The coefficient of variation of the feature x_j represents the ratio of the standard deviation to the mean, i.e.,

$$CV_j = \frac{\sigma_j}{\mu_j} \quad (10.5)$$

The difference between the maximum and minimum of all the CV values is denoted by R, and it is defined as

$$R = max\{\frac{\sigma_j}{\mu_j}; 1 \leq j \leq m\} - min\{\frac{\sigma_j}{\mu_j}; 1 \leq j \leq m\} \quad (10.6)$$

If the CV of a feature is lesser than that of another feature, one can conclude that the former feature is more consistent than the latter one. The CV is then divided by R and multiplied with $(1 - w_j)$. We use the logarithmic transformation $-2log_e(w_j^*)$ for modification of w_j. By using this transformation and equations (10.5) & (10.6), the modification of the weight function will be derived as

$$\frac{CV_j}{R}(1-w_j) = -2log_e(w_j^*) \Rightarrow w_j^* = exp[-\frac{1}{2}\frac{CV_j}{R}(1-w_j)] \quad (10.7)$$

$$\frac{CV_j}{R}(1-w_j) = exp[-\frac{1}{2}.\frac{\sigma_j/\mu_j}{R}(1-\frac{n_j}{N})] \quad (10.8)$$

For fixed values of n_j and N, if the coefficient of variation of x_j is exceeding than others, we consider x_j as a feature with less consistency. As a result, the numerical value of the modified weight function of x_j is always minor compared to a more consistent feature. So, the new weight of a less consistent feature is always inferior to a more consistent one. Mathematically, w_j^* will always lie between 0 and 1. When the product of the coefficient of variation and $(1 - w_j)$ tends to infinity, the modified weight function approaches its value toward zero. This happens only when w_j tends to zero, and the coefficient of variation is large enough. When the CV of x_j goes to zero and n_j takes its value around N, the new weight converges to 1. It is notable that in case the mean is negative, modulus has to be considered to make it positive so that $0 \leq w_j^* \leq 1$, for all j. Suppose we call it by C_{weight}.

The new weight function will help us to select the more consistent features and order them according to their significance in the model. Components with zero weightage are eliminated in the first step. Also, there will be features with less consistency, and smaller weightage dropped off from the model in the other iteration process.

Let us consider an ordered list of features $X^* = \{x_{(1)}, x_{(2)}, ..., x_{(j)}, ..., x_{(m)}\}$ appeared at least once in the selection process such that $x_{(m)}$ gets the maximum weight and $x_{(1)}$ gets the minimum weight. Now having an ordered list of features, we can fix a proportion p such that the feature $x_{(j)}$ will be included in the final model only if the corresponding $w_j^* > p$. The algorithm can be summarized as following steps.

10.3.1 Re-sampling feature selection steps

- Step 1. If the number of features is much larger than the number of samples, the filter method is used with the relaxed cut-off values to choose the factors initially.
- Step 2. Split the dataset randomly into training and testing groups and run a wrapper algorithm for selecting the features.
- Step 3. Repeat step 2 for N number of times and store the selected factors into a dataset.
- Step 4. Calculate the new weight for each of the selected factors using equation (10.4).
- Step 5. An ordered list of features is then created according to their modified weights.

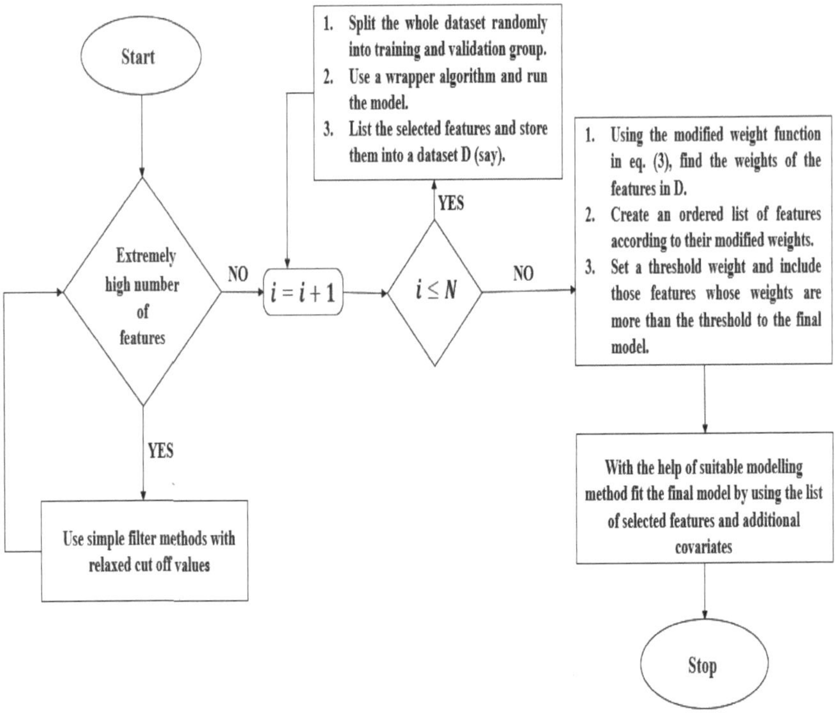

FIGURE 10.1: Flowchart of re-sampling feature selection algorithm.

- Step 6. Fix a cut-off weight and include those features which have a weight more than the cut-off value in the final model.

After fitting the model, it is always required to evaluate how well the model fits the data it was generated from. The Akaike information criterion (AIC) can help us in this context. In statistics, the AIC is used for comparison among different possible models and to determine which one is the best fit for the data [234]. This mathematical method is calculated based on the number of independent variables used for model building and the maximum likelihood estimate of the model. The AIC formula is:

$$AIC = -2lnL(\hat{\boldsymbol{\beta}} \mid X) + 2K \tag{10.9}$$

where, $L(\hat{\boldsymbol{\beta}} \mid X)$ represents the likelihood of $\hat{\boldsymbol{\beta}}$ given data X, $\hat{\boldsymbol{\beta}}$ is the maximum likelihood estimate of the parameter vector $\boldsymbol{\beta}$ given X, and K is the number of estimated parameters. The best-fit model according to AIC is the one which has minimum AIC value.

10.4 Data Methodology

This methodology is suitable for working with publicly available gene expression datasets. The dataset is obtained from Gene Expression Omnibus (GEO) (https://www.ncbi.nlm.nih.gov/geo/) with accession number GSE 65622. The dataset consists of 507 protein expression values from 80 locally advanced rectal cancer patients with multiple treatment interventions and five years follow-up duration. The relapse of the disease is our event of interest. The baseline expression data of 507 proteins is used to show the method. It is verified that a total of 28 (35%) subjects have experienced the event during the study tenure [232,233]. After downloading the dataset in CSV format, we used MS Excel to clean the dataset preliminary. R version 3.6.3 (Copyright (C) 2020 The R Foundation for Statistical Computing) is used to create the function "weightfun" (https://github.com/soutik28/Modified-Weight-Function/commit/main) and run the algorithm. This function is coupled with "glmnet," and "survival" packages in R [235].

All the proteins with 5% missing values were omitted from the data during the initial screening, and we got a total of 327 (64.5%) protein expression values. It might happen that some clinically important proteins were excluded due to the missing values since the data were used only for illustrating the method – proteins' clinical importance can be ignored. The mean of the expression values then replaces the missing values of the chosen proteins to obtain a complete dataset. Since the number of expression values was more than four times higher than the sample size, we have used the univariate Cox-proportional hazards model with a 5% level of significance for further filtering the features. The results provided us with 100 significant proteins in the first step of the algorithm. Table 10.2 shows the set of significant proteins and their corresponding p-values. Using the C_{weight} and a threshold value $p = 0.90$, we selected 12 proteins from a set of 100 to model the dataset for further analysis. These 12 proteins are the most relevant in terms of their consistency and occurrence in the iteration process. The estimated parameters corresponding

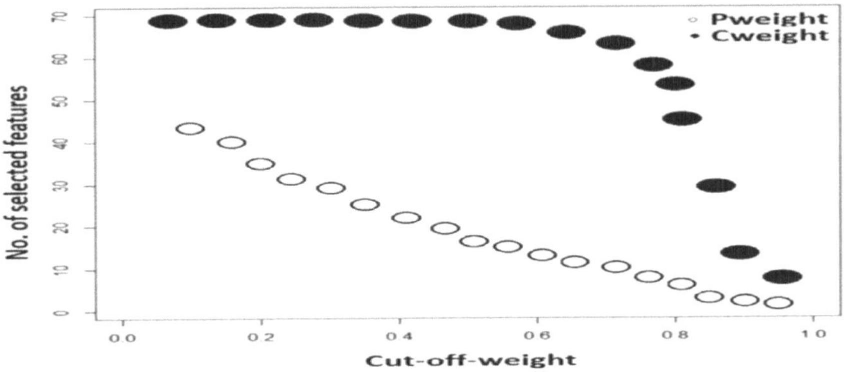

FIGURE 10.2: Number of features selected under P_{weight} and C_{weight} for different cut-off weights

to these 12 proteins shown in Table 10.3 are the $\hat{\beta}_j\{j = 1, 2, ..., 12\}$ values for the final multivariate Cox-proportional hazards model. Table 10.4 compares the feature selection of the re-sampling method using C_{weight} with the same method using P_{weight}. We get an AIC of 120.22 for the model on C_{weight} whereas the AIC value for the model on P_{weight} is 165.45. The predictive model on C_{weight} calculates the estimated hazard rate of relapse, and lower AIC represents a better model than the one which uses P_{weight}. Also, figure 10.2 shows that C_{weight} can select more features than the P_{weight} at the same cut-off weight. Therefore, a new feature selection method from a large dataset to run a predictive model is illustrated. Proteins are selected based on level of significance of 0.05 in Table 10.1.

The procedure on re-sampling technique shows a modified feature selection approach from high-dimensional data to handle any predictive model. The given example illustrates the usefulness of the limited weight function to develop a good model by choosing only 12 proteins from a set of 507. In this method, the training and validation datasets are randomly allocated once. As a result, we can overcome the biasedness owing to this random allocation only a single time.

TABLE 10.1: List of significant proteins based on level of significance 0.05.

Number	Name of proteins	P-value	Number	Name of proteins	P-value
1	Activin A	0.0429	51	Activin RIIA	0.0245
2	Adiponectin Acrp30	0.0302	52	AgRP	0.0150
3	Angiogenin	0.0009	53	BCMA TNFRSF17	0.0044
4	BMP 7	0.0321	54	BTC	0.0337
5	Cardiotrophin 1 CT 1	0.0124	55	CCR8	0.0205
6	CCR9	0.0015	56	CD14	0.0117
7	CD27 TNFRSF7	0.0001	57	Chordin Like 2	0.0042
8	CRTH 2	0.0058	58	CTACK CCL27	0.0007
9	CXCR4 fusin	0.0007	59	Decorin	0.0046
10	EGF R ErbB1	0.0022	60	Endoglin CD105	0.0026
11	Endothelin	0.0090	61	Eotaxin 3 CCL26	0.0328
12	FADD	0.0025	62	FGF 13 1B	0.0000
13	FGF 5	0.0091	63	FGF 6	0.0167
14	FGF 9	0.0243	64	Follistatin	0.0305
15	Frizzled 4	0.0000	65	Galectin 3	0.0275
16	GASP 2 WFIKKN	0.0048	66	GDF11	0.0006
17	GRO	0.0002	67	HCC 4 CCL16	0.0006
18	Hepassocin	0.0110	68	HGF	0.0072
19	HRG alpha	0.0306	69	ICAM 1	0.0388
20	ICAM 2	0.0005	70	IFN Alpha-Beta R1	0.0030
21	IFN Alpha-Beta R2	0.0215	71	IFN-Gamma	0.0012
22	IGFBP rp1 IGFBP 7	0.0306	72	IL 1 R4 ST2	0.0000
23	IL 1 ra	0.0250	73	IL 1 sRII	0.0063
24	IL 10R Alpha	0.0119	74	IL-11	0.0408
25	IL 12-p70	0.0069	75	IL-13 R-Alpha 2	0.0114

TABLE 10.2: List of significant proteins based on level of significance 0.05.

Number	Name of Proteins	P-value	Number	Name of Proteins	P-value
26	IL-17RC	0.0008	76	IL 22 R	0.0366
27	IL 28A	0.0106	77	IL-5 R-Alpha	0.0000
28	IL-7	0.0117	78	Insulysin IDE	0.0214
29	IP 10	0.0482	79	Kremen 1	0.0076
30	Kremen 2	0.0046	80	L Selectin CD62L	0.0295
31	Lck	0.0192	81	Lipocalin 2	0.0130
32	LRP 1	0.0039	82	MCP 3	0.0080
33	MFRP	0.0187	83	MIF	0.0008
34	MIP 3 beta	0.0403	84	MMP 1	0.0010
35	MMP 14	0.0371	85	MMP 15	0.0475
36	MMP 20	0.0000	86	MSP alpha Chain	0.0000
37	NAP 2	0.0059	87	NCAM 1 CD56	0.0445
38	NOV CCN3	0.0448	88	Orexin A	0.0242
39	Orexin B	0.0097	89	OSM	0.0010
40	Osteocrin	0.0002	90	OX40 Ligand TNFSF4	0.0304
41	PD ECGF	0.0052	91	PDGF R beta	0.0363
42	PDGF BB	0.0423	92	PDGF D	0.0128
43	PF4 CXCL4	0.0009	93	PLUNC	0.0101
44	SDF 1 CXCL12	0.0322	94	Siglec 5 CD170	0.0451
45	Siglec 9	0.0052	95	SPARC	0.0030
46	Thrombospondin 1	0.0195	96	Thrombospondin 2	0.0098
47	TIMP 2	0.0002	97	TIMP 3	0.0268
48	TMEFF1 Tomoregulin 1	0.0000	98	TNF RII TNFRSF1B	0.0138
49	TNF alpha	0.0023	99	TRAIL R1 DR4 TNFRSF10A	0.0357
50	TRAIL R3 TNFRSF10C	0.0134	100	TRANCE	0.0286

TABLE 10.3: Proteins list with their weights and CV after 150 iterations using threshold 0.90.

Number	Name of Proteins	$P_{weight}(w_j)$	Coefficient of Variation	$C_{weight}(w_j^*)$	$\hat{\beta}_j$
1	FGF 5	1.00	0.39	1.00	0.0101
2	GASP 2 WFIKKN	0.95	0.41	0.99	0.0047
3	MSP Alpha Chain	0.94	0.52	0.98	0.0011
4	Trail R3 TNFRSF10C	0.87	0.39	0.98	-0.0681
5	L Selectin CD62L	0.84	0.32	0.97	-0.0089
6	ICAM 1	0.85	0.41	0.97	0.0314
7	PDGF BB	0.82	0.50	0.95	-0.0062
8	EOTAXIN 3 CCL26	0.83	0.72	0.94	-0.0045
9	PDGF R Beta	0.72	0.59	0.92	0.0033
10	IFN Gamma	0.59	0.43	0.91	-0.0003
11	Engoglin CD105	0.58	0.44	0.91	0.0000
12	FGF 13 1B	0.62	0.50	0.91	0.0004

TABLE 10.4: Number of selected features and AIC values.

Re-sampling method	Number of features	AIC
C_{weight}	12	120.22
P_{weight}	3	165.45

10.5 Weight Function and The Re-sampling Algorithm

The training datasets play an essential role in any feature selection method. It is investigated that embedded methods enormously rely on the taste of the training set and choose the features automatically based on some regularization functions [236, 237]. Moreover, the re-sampling algorithm does not like any elements from the training set by default but makes an ordering of the features by their relevance in the model [233]. The parts are then adjusted and selected in the ultimate model by choosing a threshold for the modified weight w_j^*. Therefore, we can observe that the C_{weight} function can eliminate less informative proteins more consistently than the previously used P_{weight} function, as stated in [233].

A current study on high-dimensional data analysis uses the Bayesian concept on a group of generalized ridge regularization. This approach modifies the shrinkage using prior information of the regression coefficients [238]. On the contrary, the derived re-sampling algorithm does not choose any variables according to their estimate of the parameters. Before fitting the model, we first identify the variables based on the modified weight. And then, we only take the relevant components to the ultimate model. Hence, one can perform manual incorporation of any additional variables in the final model by relaxing the cut-off value.

In clinical research, sometimes the investigator prefers to control what is needed to include in the ultimate model than what is chosen by the method. The re-sampling process and the modified weight function give the investigator more control in developing the model. However, the choice of training set influences the final group of variables in the existing embedded methods [239]. Those methods may choose a different list of variables if we use an additional training or testing dataset or change the random sample proportions. The re-sampling algorithm and the modified weight function can now resolve this issue.

In the application of microbiome data, the goal is to identify the actual underlying biological signal. Reproducibility criteria for evaluating feature selection methods, such as Nogueira's Stability measurement, can help to determine the optimal feature selection method in this context [240, 241]. In contrast, the proposed method removes the matter of lacking reproducibility in the random selection of training and testing groups by considering the consistency of the variables and sorting criteria based on the steps of the algorithm. Recent advancement in genomics is bio-marker modeling that uses exquisite computational modeling methods like joint longitudinal modeling, Bayesian state-space modeling, time to event modeling, etc. [242–244]. One can treat the re-sampling form and the modified weight function to identify the candidate features for these models. The prediction accuracy of these models can be improved by the re-sampling method. Furthermore, we require research on these advanced predictive models to measure the influence of different feature selection algorithms.

The above results display that the re-sampling method can produce a compelling selection of variables based on the new weight function. Still, we need a high programming computer and an amount of significant time due to re-sampling. Any further study on this method can evaluate its effect to solve other biases existing in microarray feature selection methods, such as multiple testing, imbalance class, normalization, etc. [239, 245].

There are some limitations of the proposed approach. It is known that the coefficient of variation can be calculated only for the variables. It is either discrete or continuous, not for the attributes. In this consequence, the proposed re-sampling strategy does not take care of the categorical type variables. Since we couldn't calculate the CV for any features, the construction of the new weight function is improper for these types of factors. The choice of cut-off weight is another limitation of the proposed algorithm. Specification of the threshold is always in the hand of the researcher, and its proper selection can lead to building a good prediction model and decreasing the AIC value.

There are different aspects in the analysis of cancer relapse. The essential task in the clinical setting is to supervize relapse-free survival and related outcomes. The time to relapse is the outcome variable in those similar situations. It will be more complicated to analyze the problems in the presence of other competing events such as death [246]. We can also use our algorithm for analyzing the time to event outcome by choosing suitable models like the logistic model and non-proportional hazards model. Assessment of the re-sampling algorithm together with the modified weight function under the different model buildings can be a topic for future research.

Illustration of the modified method on weight function and the re-sampling algorithm provides us with a new methodological way to select consistent features from a large dataset. This article also compares the new process and the existing one, which helps a researcher perform the predictive model. In addition, the example exhibits the usefulness of the proposed approach for developing an accurate model with a low AIC value by selecting only 12 out of 507 proteins. This new approach will help us overcome the challenges of biasedness in determining training and testing sets. However, we have required adequate time and sound computer systems to run the algorithm properly for a high-dimensional dataset. The obtained results from the new approach still interpret the solutions more practical and provide enough space for the researcher to interact and choose features according to their importance.

10.6 High Dimensional Time to event

In this age of machine learning techniques, dimensional reduction and pattern recognization of the data are widely adopted through data mining techniques. But dimension reduction is always performed by looking at the outcome of interest. It becomes tedious when the outcomes of interest are measured with time to event data. However, there is an existing gap in utilizing machine learning techniques in high-dimensional survival data due to the non-availability of convenient programming functions and packages. This chapter is dedicated to applying the machine learning approach to high-dimensional data reduction techniques. The outcome variable is presented as time to event. Few machine learning packages are suitable to work with high-dimensional data. But these are not compatible with time to event data analysis. The package developed to work with high dimensional time to event data is presented in this chapter. A high-dimensional gene expression dataset has been analyzed using the proposed R library to show its efficacy in feature selection.

10.6.1 Time to event data

In survival analysis, the key focus remains on predicting the time to a specific event of interest with a certain accuracy limit. The prediction of time-to-event helps in identifying a particular patient or a group of patients with increased risk of disease occurrence [247]. The predictive models based on time-to-event outcomes help make decisions regarding suitable treatment strategies. A critical aspect of survival data is that it is often censored, which happens when the particular event is not observed within the pre-scheduled duration [191]. Due to the presence of censoring, the analysis of survival data is usually more complicated than the regular statistical analysis [248]. Different types of parametric (linear regression, accelerated failure time), semi-parametric (Cox proportional hazard (CPH) model), and non-parametric (Nelson-Allen, Kaplan Meier (KM)) methods have been developed to date for the analysis of survival data and enable us to answer more intricate questions regarding disease prognosis [249]. The rapid advances in medical science encourage more sophisticated modeling techniques of survival data. Continuous efforts have been made to find out finer tools for the detection of disease progression [140]. This necessitates the introduction of better biological indicators. In this aspect, genomic profiling of patients gives an edge in understanding the disease progression and prognosis, which aids in discovering a better treatment strategy [250].

10.6.2 Gene expression data

In genomic profiling, expression is collected from the patients for a large number of genes. It is a method to understand the genetic information pertaining to an individual person or specific cell types and also the interaction between genes [251]. However, a high number of gene expressions complicates the prognostic modeling. Due to the analogous biological characteristics of multiple genes, we obtain very little gain in information by including an excessive number of genomic profiles. A typical problem related to this type of expression dataset is that many of the gene features are either highly correlated or unnecessary [252]. Also, the study becomes computationally challenging to incorporate all the gene profiles, and the actual contribution of significant genes is often reduced. So, it requires some statistical tools to reduce the curse of dimensionality for more efficient analysis [253].

The problem of dimensionality reduction and classification has become a key interest in several research problems, and numerous methods have been addressed in this regard. This is where machine learning techniques prove their necessity.

10.6.3 Machine learning algorithms

Machine learning algorithms are gaining popularity in predicting the survival outcomes of patients. In parametric machine learning techniques like linear regression, logistic regression [253], naive Bayes algorithm [254], a finite number of parameters are assumed for training the model [255]. One of the key advantages of using non-parametric machine learning methods is that they do not assume any prior knowledge of the data generation process and are able to include several types of data learning processes like neural networks, radial basis function (RBF), kernel support vector machines (SVM), bagging and boosting, decision trees, etc. [256]. Several instances are available where machine learning algorithms have been successfully used for survival prediction [257,258]. There are numerous uses of ensemble learning approaches that can combine several machine learning algorithms for the prediction of clinical outcomes. An overview of these techniques has been discussed in the following survey article [259].

With the development of high throughput technologies, more and more high-dimensional genomic data are generated. In genomic profiles, the number of covariates (p) (i.e., gene features) is usually much larger than the sample size (n) (i.e., $p >> n$) [259]. This leads to a challenging problem of dimension reduction. The primary focus of this article is on the analysis of high-dimensional gene expression data that occurs in oncology studies. With the rapid advancement of clinical decision-making for more precise treatment, gene expression of different genomic profiles has shown potential for the efficient prognosis of the disease. It is required to apply the existing statistical methods so that necessary information can be easily deduced. One of the essential aspects of oncology research is the classification of biomarkers according to their effect on disease progression [260]. Identification of changes in gene expression values at different time points can provide new insight into the discovery of effective biomarkers and also disease-based classification of genes. Often it is found that the different levels of gene expression values have varying effects on disease progression [261]. For example, the cancer progression risk with HER2 positive status differs from the same with HER2 negative status. This influences the corresponding medical decision-making for more suitable treatment. At the same time, the survival duration of patients with the same biomarker level is likely to be correlated. So, our interest is to find the optimal threshold levels in gene expressions which will assist in identifying the heterogeneity in disease progression risk. Complexity appears when several biomarkers possess co-dependency that affects the corresponding expression values [256]. This type of problem for threshold selection of co-dependent biomarkers demands a sophisticated computation tool for medical decision making which can be both flexible and efficient to perform necessary analysis in one go.

10.6.4 Machine learning codes with high dimensional data

Machine learning can review large volumes of data and produce concise predictions by discovering specific patterns in the dataset without any human intervention. With the use of regularization, it updates the learning algorithm on its own and produces a simpler model with a high degree of predictive accuracy. However, the development of statistical software capable of applying machine learning algorithms in survival analysis lags far behind. There is a clear scarcity of the related programming codes has been observed in this area. The R package **survival** [53] is capable of incorporating left, interval, and right-censored data along with competing risk, multistate models, and model evaluation. However, it only supports classical statistical models and not any machine learning approaches. The R package **mboost** [262] uses a functional gradient descent algorithm for optimizing the risk functions. It uses least square estimates or regression trees as a base learner model for fitting different models to potentially high-dimensional data.

Package **RegCox** in R and **TransferCox** in C++ is capable of applying active learning and transfer learning on survival objects [259]. The package **MLSurvival** can train and evaluate machine learning survival and classification models for time-to-event data. The R library named **mlr3proba** introduces survival modeling using machine learning approaches under the **mlr3** package [263]. It allows for multiple survival analysis predictions and selects the best features using specialized survival measures. The focus of this paper is to deliver a software package that will be able to perform machine learning algorithms on clinical data with the assembly of survival objects from the survival package and produce the desired number of features.

Here the package **highMLR** delivers a unique solution for the analysis of high dimensional gene expression data, classification of genes according to their effect on disease, and identification of death and disease progression risk heterogeneity in gene expressions. The package takes into account high dimensional gene expression data and performs feature selection based on a specified learner method. Currently, there are two learner methods available, namely the CPH model and the KM method. For each of the genes, the data is cross-validated by splitting it into the training set and test set. With the specified learner method, it builds a model by implementing it on the training dataset and validates the model on the test set. A performance measure is required for the comparison of the model fitting on different genes for which we considered the logarithmic loss function. The package **highMLR** provides an implementation of this feature selection technique with the functions `mlhighCox` and `mlhighKap`.

This package is capable of performing frailty analysis on the selected genes with the options of several distributions to identify the heterogeneity in the expression values. For the biomarker studies, the identification of effective levels of gene expression demands significant attention for disease prognosis. Through the function `mlhighHet`, our package provides this advantage for medical practitioners to identify the risk heterogeneity by measuring the randomness involved at different levels of the expression values.

It uses a sequential CPH approach using a minimum p-value and chooses the optimal threshold level for each gene. The package also provides a provision of marker classification along with feature selection through the functions `mlclassCox` and `mlclassKap`.

Here we demonstrated analysis of a high dimensional survival dataset using the **highMLR** package. The application of this package is explained on two protein marker datasets to show its efficacy.

10.7 Methodological Framework

10.7.1 Feature selection

A necessary aspect of gene expression studies is to identify the most significant genes from a large pool of genomic profiles. The "curse of dimensionality makes it difficult to classify the genes and prognosis of disease onset. So, the feature selection mechanism provides its essentiality to choose only a handful of gene biomarkers based on their relevance. In these methods, the learning models are optimized using some pre-specified measures and then tuned to the randomness of the input gene expression data to form the clusters.

A likelihood-based gene selection method was proposed by Abdullah, whose performance was verified on a cancer dataset [264], and a feature elimination method using a machine learning method [214]. Using the weights of trained SVMs, the algorithm eliminates the genes according to a less significant contribution to the classification. We prepared the dataspace as

$$D = \{(x^1, y^1), ..., (x^n, y^n)\} \subset (X, Y)^n \tag{10.10}$$

where X is the input space with $p = dim(X)$, Y is the output or target space, the tuple (x^i, y^i) denotes the ith observation. Here, $x_j = (x_j^1, ..., x_j^n)$ denotes the jth feature vector. So, we have a complete data structure with n objects and p feature vectors and we have to choose $r(r < p)$ most significant features. In most cases, the true nature of the data pattern is unknown. A model f is chosen, which maps the feature vectors to the target variable.

The selection of the learning model should be based on the nature of the dataset, which can capture the intrinsic relationship between target and input variables. Identifying the optimal model is equivalent to choosing the optimal set of parameter values. With the detection of optimal parameters, we fit the data into the functional relationships that map the object's features into the target.

The algorithm for finding the most effective model f is called learner. It is a mapping from the data space with the vector of hyper-parameters to a model. In learners, the model space is defined in advance. Using the training dataset and hyper-parameters, the learner algorithm finds the parameters that make the model most suitable for that dataset. In this study, two different types of learner models are considered for the survival objects, namely, CPH model and KM method. The CPH model is most popularly used for survival analysis. With the standard definition of failure time and event indicator for a survival object, the CPH is expressed as,

$$h_i(t|K_i) = h_0(t)exp(\mathbf{K_i}\gamma)/\mathbf{5} \tag{10.11}$$

where $h_i(t|K_i)$ is the hazard function for the ith subject who experiences the event of interest at time t, $h_0(t)$ is the baseline hazard function. The baseline covariates $\mathbf{K_i}$ with corresponding regression coefficient is presented as γ. The Kaplan-Meier estimator of survival function $S(t)$ is defined as,

$$\hat{S}(t) = \prod_{i:t_i \leq t} \left(1 - \frac{d_i}{n_i}\right) \tag{10.12}$$

with t_i is the time with at least one event happened, d_i is the number of events happened at time t_i, n_i is the number of individuals at risk up to time t_i. Here, it is assumed that the survival probabilities are same for all the subjects irrespective of their entry to the study.

To obtain the best model to fit the dataset, we need to verify it with data where the information of the target variable is unknown. However, obtaining a completely new dataset is not always possible and often demands a substantial amount of effort. So, to assess the performance of a learning algorithm, a statistical technique popularly used for validation purposes is known as Cross-Validation. Here, the data is split into k different and roughly equal parts. At each time, one part is separated as a test set, and the rest $k-1$ of them are used as the training set.

The model is built based on the training set using the learner model, and the resulting model is fitted to the test set. The predicted values are then compared with the true target values, and the performance of the model is verified through the distance between predicted values and true values.

Several performance measures in the form of loss function are available to check the model's accuracy in predicting the target variable. In this study, we used the logarithmic loss function in the context of probabilistic prediction for survival objects. The log loss function is defined as the negative log-likelihood density evaluated at time t. It is expressed as, $L(f,t) = -log(f(t))$.

The standard error of the log loss L is approximated as $s.e.(L) = sd(L)/\sqrt{N}$ where N is defined as the number of subjects in the test set. At each step of the cross-validation technique, the logarithmic loss function is calculated, and the result is aggregated to get an overall measure of performance.

The lower value of the log loss function implies a better likelihood value and superiority of the model. Based on the minimum log loss values, the desired

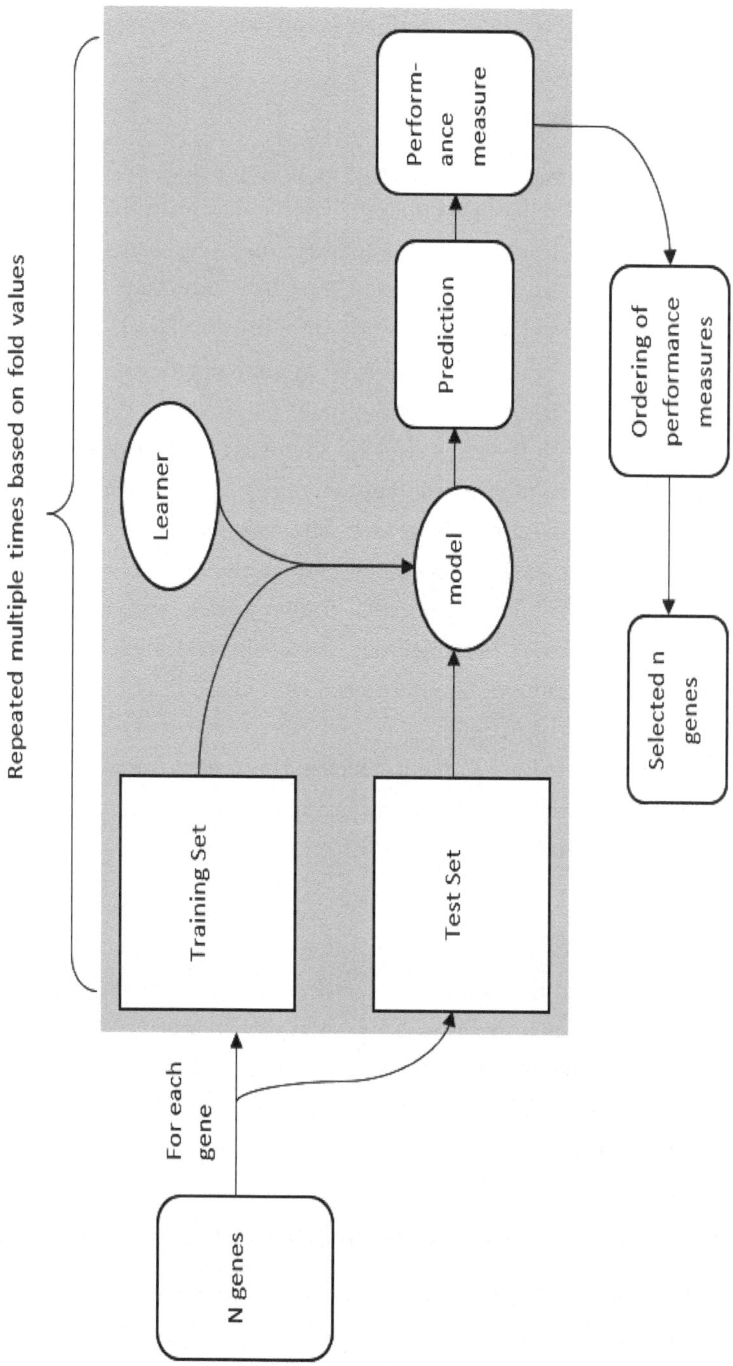

FIGURE 10.3: Feature selection algorithm.

number of features are selected from the high-dimensional survival dataset. The feature selection algorithm is described graphically.

10.7.2 Frailty analysis

The presence of disease risk heterogeneity is an important criterion in gene expression studies. Researchers have found that the survival probability of a patient depends on the levels of the biomarker, and the treatment regime is also needed to be planned accordingly. Thus, it is necessary to identify the heterogeneity in expression values among the subjects. This heterogeneity between the cluster of observations which is often unobserved, can be modeled using a random frailty term.

The term frailty was first introduced by Vaupel [126] and discussed the univariate frailty model to address the unobservable random effect shared by subjects in a standard-risk group. Keyfitz & Littman [265] formulated a procedure for the estimation of life expectancy by considering the heterogeneity present in the population. Several possible frailty models were analyzed to incorporate different types of heterogeneity in the survival model [266]. The variance of the frailty term can be interpreted as a measure of heterogeneity between different groups of patients.

We have considered the CPH model, where the frailty component is incorporated multiplicatively with the covariate information. The CPH frailty component with the standard definition of components defined above is expressed as [49],

$$h_i(t|K_i) = h_0(t)z_i exp(K_i\gamma) \tag{10.13}$$

where z_i is the shared frailty component with $var(z_i) = \nu$. Several distributional assumptions like gamma, log-normal, and exponential have been considered for the frailty component to assess different forms of heterogeneity structure in the subjects. The estimation of parameters is performed using penalized likelihood framework.

10.7.3 Classification using CPH model in time-course data

The CPH model has been successfully utilized to identify the optimal threshold level of a single biomarker. The survival duration of a patient is divided into a few parts based on the number of follow-ups. We assume that the number of follow-ups for patient i is l_i. The maximum value of l_i will be J, a pre-defined value for the study (in our study, the value of J is 4). The follow-up durations are $t_1, t_2, ..., t_{l_i}$. So, the follow-up times are $0, t_1, t_1+t_2, \ldots, t_1+t_2+...+t_{l_i}$. The gene expressions are assigned according to the above-defined procedure. The intervals are assumed to be independent.

For each time point j and each gene g a CPH model is estimated considering the fixed gene expression value of gene g during the jth duration period. More precisely, for i^{th} individual, the proportional hazard model fitted at time point j for a particular gene (g) can be written in simple terms as

$$\lambda_{ij}(t|Z_{g,ij}) = \lambda_0(t)exp(Z_{g,ij}\beta_g) \tag{10.14}$$

or,

$$\log\,\lambda_{ij}(t|Z_{g,ij}) = \log\,\lambda_0(t) + Z_{g,ij}\beta_g, \quad j = 1, 2, ..., J \tag{10.15}$$

where $Z_{g,ij}$ is an indicator function defined as follows. For each gene g at time point j, the expression values are divided into deciles and we have to find the optimal threshold value of activity of these genes. The choice of decile is not unique.

There are several facts like number of data-points, scientific knowledge and the judgment to decide the threshold. For more refined output of the threshold values, one can always choose other percentiles of the expression values. For a particular p^{th} decile [p =1,2,...,9], the expression is divided into two parts, $[min\ Y_{g,j}, Y_{(g,j)p}]$ and $[Y_{(g,j)p}, max\ Y_{g,j}]$. $Y_{g,j}$ denotes the expression value for gene g at time-point j, $Y_{g,ij}$ denotes the gene expression value for i^{th} patient at j^{th} timepoint, $Y_{(g)p}$ denotes the pth decile of expression value for gene g, $Y_{(g,j)p}$ denotes the pth decile of expression value for gene g at j^{th} time point. Thus, the indicated $Z_{(g,ij)p}$($Z_{(g,ij)p}$ is the form of $Z_{g,ij}$ for p^{th} decile) is defined by, $Z_{(g,ij)p} = \begin{cases} 0 & \text{for } Y_{g,ij} < Y_{(g,j)p} \\ 1 & \text{for } Y_{g,ij} \geq Y_{(g,j)p} \end{cases}$. In this way, the CPH models that are fitted for different indicators on time-point j for gene g are,

$$\log\,\lambda_{ij}(t) = \log\,\lambda_0(t) + I(Y_{g,ij} > Y_{(g,j)p})\beta_{(g,j)p} \tag{10.16}$$

or,

$$\log\,\lambda_{ij}(t) = \log\,\lambda_0(t) + Z_{(g,ij)p}\beta_{(g,j)p}, \quad p = 1, 2, ..., 9 \tag{10.17}$$

where $\beta_{(g,j)p}$ is the corresponding coefficient for p^{th} decile at j^{th} time-point. For each equation, we obtain a corresponding p-value that demonstrates the significance of the covariate. The lower the p-value, the greater the significance of the covariate on response.

Here, the equation with a minimum p-value gives the highest importance to the indicator, which produces the highest significant level of gene expression. Thus, for each time point, we obtain the threshold values of gene expression. Note that the proposed procedure for defining the best threshold of each gene at each time point using the minimum p-value procedure does not correct the significance level of the best threshold selected.

The correction of the significance level for a series of several codings using the minimum p-value procedure has been addressed for the generalized linear model and Cox regression. In our study, the minimum p-value procedure is used as a step to allow us to find the best cutoff for each gene at each time point in order to find potential biomarkers. The classification of genes is performed according to their effect on the disease progression at each time point.

The gene expressions are assumed to be independent of each other. Gene expressions of some particular genes increase with the disease occurrence, while for some genes, the disease negatively affects expressions. So it is important to take into account the signs of the parameters associated with the indicator of threshold values. For the gth gene at J time-points, the parameters obtained using CPH model are $[\beta_{(g,1)}, \beta_{(g,2)}, ..., \beta_{(g,J)}]$. If $\beta_{(g,j)} > 0$ for all j, then the gene is defined as *Progressive Gene* and if $\beta_{(g,j)} < 0$ for all j, then the gene is defined as *Oppressive Gene*. Other than these two types of genes, there are some other genes that do not have a consistent positive or negative association with the disease occurrence over time, i.e., $\beta_{(g,j)}$ is neither > 0 nor < 0 for all j. These genes are defined as *Volatile Genes*.

Once the gene classes are formed, the desired number of genes can be selected from each class to form a prognostic score which can forecast the survival outcome of the patients with the required information of gene expression values [260].

10.7.4 Sequential threshold selection

Different studies have shown that there is a significant association among the biomarkers related to a particular cancer type. Due to the complex co-dependency structure in expression values among the biomarkers, identifying optimal threshold values of gene expressions is a challenging job as the expression values of one marker influence the other. So, to overcome this issue, one can take different threshold levels of multiple gene expressions and check for the optimal threshold levels corresponding to the minimum p-value using the CPH model.

However, this approach is time-consuming and suffers the problem of multiple testing, which increases the chance of false positives. In order to overcome this constraint, we used a sequential computation approach to perform this analysis and obtain the optimal threshold levels. All possible sequences of featured genes are tabulated in advance, e.g., if n genes are selected after feature selection, we get $n!$ gene sequences.

The flowchart of re-sampling feature selection algorithm is in Figure 10.1. The number of features selected under weigted methods are given in Figure 10.2.Feature selection algorithm is presented in Figure 10.3.Figure 10.4 present the flowchart of the mlhighHet function. Longitudinal trajectory of 2 selected biomarkers in mlclassCox function is presented in Figure 10.6. Let, the set $(X_{(1)}, X_{(2)}, ..., X_{(n)})$ denotes a particular sequence of n biomarkers. The expression values are standardized to put those variables under the same scale and more suitable for comparison. A CPH model with all the featured genes is computed where, keeping aside the first gene, the rest $(n-1)$ gene expressions are transformed into a binary variable with respect to the median value. Now, based on the percentiles of the expression values, the first gene is transformed into a binary variable and incorporated into the model, as explained in equations 10.7 and 10.8. So, the CPH model becomes,

$$\lambda(t|Z_{1,p}, Z_{2,50}, ..., Z_{n,50}) = \lambda_0(t)exp(Z_{1,p}\beta_1 + Z_{2,50}, \beta_2 + ... + Z_{n,50}\beta_n) \quad (10.18)$$

Here, $Z_2, ..., Z_n$ are indicator variables formed based on the median values of those expressions. For each quantile p, the corresponding p-value of the β_1 is computed. The value of p for which the corresponding p-value of β_1 is minimum provides the optimal threshold level of that biomarker.

Using the minimum p-value approach, we obtain the optimal threshold level of that gene for which the level of heterogeneity is maximum. The threshold level with a minimum p-value gives maximum importance to each partition which will also provide minimum frailty variance. Let that optimal quantile for the first biomarker is p'.

After the detection of p' for the first biomarker, the same process is continued for the next biomarker, where the first marker is transformed into an indicator variable according to the obtained p'^{th} quantile. The CPH model for 2nd marker can be expressed as,

$$\lambda(t|Z_{1,p'}, Z_{2,p}, Z_{3,50}, ..., Z_{n,50}) = \lambda_0(t)exp(Z_{1,p'}\beta_1 + Z_{2,p}, \beta_2 + Z_{3,50} + ... + Z_{n,50}\beta_n) \tag{10.19}$$

and following the same minimum p-value procedure, we obtain the value of p for which the partition is optimal.

Thus, at the end of the whole process, we obtain a sequence of optimal threshold values of the biomarkers. However, due to the correlated nature of the biomarkers, this process is order-dependent, i.e., it depends on the order of the genes by which the optimal values are calculated.

So, the CPH models with optimal threshold levels are compared using Bayesian Information Criterion (BIC) values which are defined as $k\ ln(n) - 2ln(\hat{L})$. The minimum BIC value directs the model with optimal threshold levels of the genes. The model with minimum BIC value was used to find the most consistent levels of the expression values using frailty variances.

10.8 llustration Using R

The R package **highMLR** has been developed to implement a feature selection algorithm in high-dimensional survival data and perform a number of survival analyses on the selected features. The primary focus of this package is toward the gene expression dataset obtained from different oncology studies, where only a few features are required to be selected from a large pool of markers for efficient analysis.

In particular, this package performs feature selection of survival data using the package **mlr3proba**. It takes the survival objects using `Surv` function of **survival** package into a single object capable of handling the right censoring. The package **highMLR** is capable of handling two types of learner models namely, `coxph` and `kaplan` method. With the specified number of `fold` values, it generates test data and training data and performs cross-validation techniques to evaluate the performance of learning techniques. Based on minimum negative log-likelihood values, it selects the desired number of features.

The package **highMLR** is capable of classifying genes according to their effect of disease progression using a CPH model. The optimal threshold levels of gene expressions are identified at each longitudinal time-point using the categorization method used in Section 2.3. A sequential CPH approach is adopted to obtain the optimal threshold levels of the dependent markers. A detailed description of the functions is documented in the online help file of the package.

10.8.1 Implementation details

As an illustrative example of the functions mentioned in the **highMLR** package, we have used two different datasets of biomarkers. The first dataset is a high dimensional gene expression dataset of head and neck cancer with survival information of 565 patients. There are expression data available for 104 markers (`GJB1,...,HMGCS2`) along with the information of overall survival duration (`OS`), survival event (`Death`), progression-free survival (`PFS`) and event of progression (`Prog`).

This dataset is provided in the package entitled `hnscc`. The second dataset is a high-dimensional time-varying expression data of protein markers. The dataset contains longitudinal information of 246 markers for 79 patients with overall survival duration (`OS`), survival event (`event`), and visit number (`Visit`). More details of this dataset can be found in the Gene Expression Omnibus (GEO) database under the accession number GSE65622.

10.8.1.1 Feature selection using CPH learner model

In the initial description of the dataset, we can find that there is a large number of protein-marker information associated with the study, and it is quite difficult to analyze incorporating all the markers in the prognostic model. So, a feature selection is necessary to trim the dataset taking into account the survival information of the subjects. We start by loading the **highMLR** package and loading the `hnscc` dataset.

Listing 10.1: Call HighMLR package in R

```
> library(highMLR)
> data(hnscc)
```

The following line of commands will perform a feature selection method by CPH as a learner model. The `cols` define the columns to be entered in the selection process, `per` denotes the percentage of features to be selected and `fold` defines the number of folds required for the cross-validation. The default fold value and percentage of features are 3 and 20, respectively. We selected 10

Listing 10.2: Extracts desired number of features based on minimum logLoss function using Cox proportional hazard model as learner method on a high dimensional survival data

```
> test = mlhighCox(cols=c(6:15), idSurv=``OS",
idEvent=``Death", per=20,fold = 3, data=hnscc)
> print(test)
   gene nr          resample_result task_id learner_id
1: RGL3  1 <ResampleResult[21]>       data surv.coxph
2: GJB1  1 <ResampleResult[21]>       data surv.coxph
   resampling_id iters    surv.logloss
1:       cv          3    0.5429207
2:       cv          3    0.5469711
```

markers for the analysis, and from that, 20% i.e., 2 features will be selected. The output shows that the selected features are `RGL3` and `GJB1`. Due to the cross-validation technique used in the feature selection method, the output is not uniform, and with the change in fold values and number of occurrences, the output may vary. So, it is advisable to run a few iterations of the process and obtain those features with the most number of times appearing in the output.

10.8.1.2 Feature selection using kaplan method learner model

A similar process can be performed using a Kaplan method as the learner model. The commands are given below. There is a possibility of not having enough events in the training dataset during cross-validation, which will hamper the machine learning procedure. If the dataset contains an inade-

Listing 10.3: Extracts desired number of features based on minimum log-Loss function using Kaplan Meier model as learner method on a high dimensional survival data

```
> test=mlhighKap(cols=c(6:15), idSurv=``OS",
idEvent=``Death", per=20,fold = 3, data=hnscc)
> print(test)
     gene nr        resample_result task_id
learner_id resampling_id iters
1:    TOX3  1 <ResampleResult[21]>
data surv.kaplan            cv    3
2: SLC4A4  1 <ResampleResult[21]>
data surv.kaplan            cv    3
    surv.logloss
1:     0.5379926
2:     0.5454196
> test$gene
[1] ``TOX3"    ``SLC4A4"
```

quate number of survival events, in that case, the number of folds should be decreased. This will ensure that the training set consists of enough survival events for the proper application of the learner model. The `test$gene` provides the selected features and `surv.logloss` gives the value of the logarithmic loss function.

10.8.1.3 Fraity analysis with high dimensional data

Analysis of subject-specific heterogeneity can be conducted in `highMLR` package using the function `mlhighFrail`. The variance of the frailty component is computed as the measure of heterogeneity present in the dataset. The `mlhighFrail` uses the CPH model as the learner method for feature selection, and then the frailty variances are computed using a CPH frailty model. Three different distributional assumptions are considered for the frailty variable, namely, Gamma, Gaussian, and t-distribution. In the analysis of `hnscc` dataset, the variable indicating patient id (`idFrail`) is considered as the frailty component.

A table providing information on the selected markers along with frailty variances is obtained as the output.

Listing 10.4: Extracts features based on minimum log-Loss function using Cox proportional hazard using high dimensional survival data

```
> test=mlhighFrail(cols=c(10:20), idSurv=``OS",
idEvent=``Death", idFrail=
            ``ID", dist=``gaussian", per=20, fold = 3,
data=hnscc)
> print(test)
  gene frailty variance
1 HGD   0.05631532
2 LRP2  0.04166667
```

10.8.1.4 Sequential thresholding of correlated biomarkers

An important aspect of oncology study is to identify the risk heterogeneity in gene expression values. However, due to the presence of dependency among the biomarkers, it is not ideal to detect the threshold levels of the biomarkers, considering them as independent entities. The procedure works in three stages, (i) it selects the required number of features using a feature selection algorithm, (ii) using the sequential CPH model as described in Section 2.4, we obtain the optimal threshold values of the markers, and (iii) the CPH frailty model is applied for different partitions of the subjects based on those optimal threshold levels to identify the most consistent level.

For the `hnscc` dataset, we select 9 markers (columns 27 to 35) for the analysis, and our objective is to select 3 markers with their consistent levels and optimal threshold values. The commands and output described below will show the analysis.

Here we can find that the selected features are CACNA1D, PCP4, and SYT13 with their respective optimal threshold values -1.067627, -0.4952277, and 1.44947 and the corresponding BIC value of the model 2812.423. Based on the optimal threshold levels, we obtain the class of expression values with the least heterogeneity, which is CACNA1D > -1.067627, PCP4 > -0.4952277, and SYT13 < 1.44947. With respect to the markers selected and their corresponding threshold values, this class shows the most consistency in survival prediction. The risk heterogeneity for different levels of expression values of the 3 markers is shown through Kaplan Meier curves in Figure 10.5.

One assumption considered for obtaining the most consistent level is that we incorporated only those classes where at least 5 observations are available.

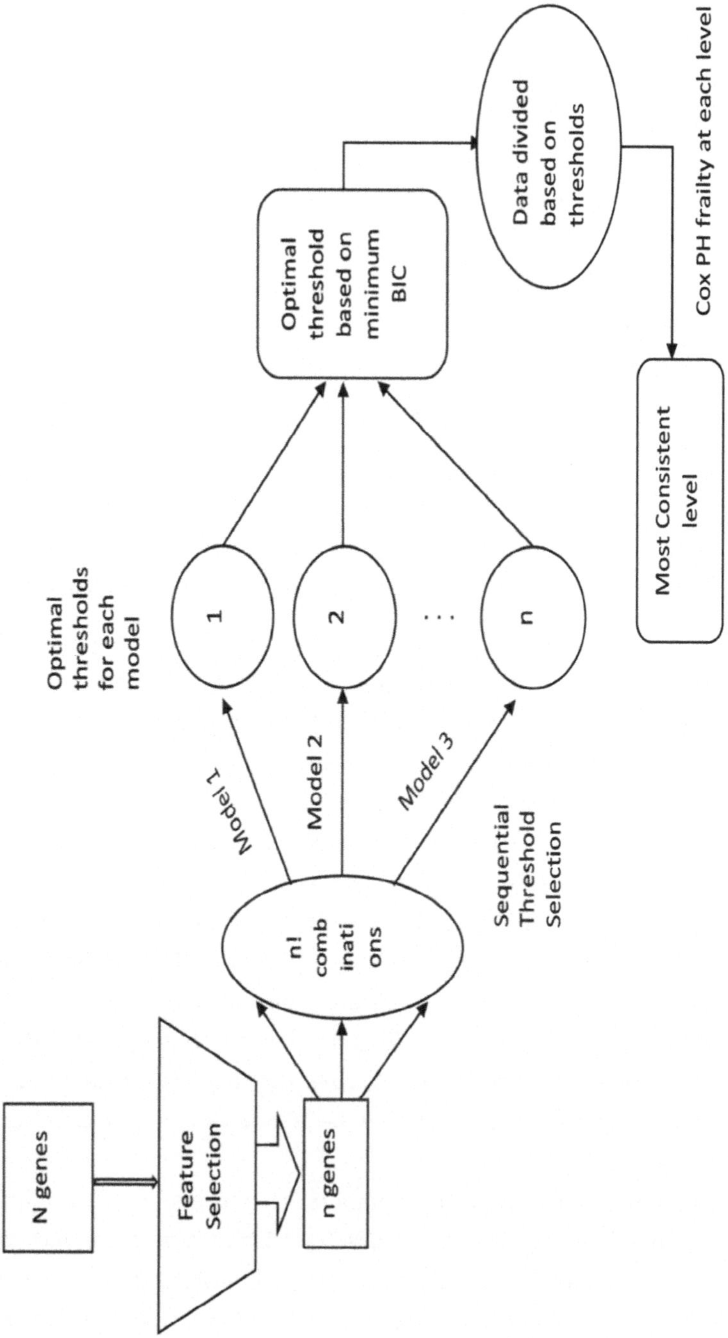

FIGURE 10.4: Flowchart of the `mlhighHet` function.

Listing 10.5: Extracts features based on ML method, finds optimal cut-off values of features using sequencial Cox PH model and obtain the most consistent level according to the cut-offs

```
> test=mlhighHet(cols=c(27:35), idSurv="OS",
idEvent="Death", idFrail="ID",
        num=3, fold = 3, data=hnscc)
> test$optimal_cutoff
      gene1 gene2 gene3 opt_cutoff_gene1
opt_cutoff_gene2 opt_cutoff_gene3
1  CACNA1D  PCP4              SYT13    BIC
-1.067627    -0.4952277      1.44947   2812.423
> test$consistency_level
      gene1 gene2 gene3 opt_cutoff_gene1
```

Otherwise, convergence issues may appear due to the lack of an adequate number of subjects.

10.8.1.5 Gene classification using longitudinal gene expressions

Classification of genes according to their effect on disease progression deserves attention for a more efficient prognosis and design of better treatment strategy for patients. In the package **highMLR**, the functions `mlclassCox` and `mlclassKap` present an efficient procedure for gene classification using the path of longitudinal movement of expression values. The `srdata` in **highMLR** package contains gene expression values of 246 markers with longitudinal trajectory and survival information. The package also gives the provision of handling missing observations in the data. If the data contains any missing value, it will run a multiple imputation works using `missForest` function of package **missForest** and replace the missing values with imputed data.

In the following commands, we will show how the function utilizes the longitudinal expression values and categorizes the markers. 20 markers are selected for the analysis, out of which 30%, i.e., 6 markers have been obtained as per the requirement.

The output shows the forest-based imputation process has been applied for the missing observations. The `test$GeneClassification` shows that there are 2 Positive genes and 4 Volatile genes obtained with no genes belonging to the class of Negative genes. The longitudinal trajectory of those 6 genes, along with their mean expression levels at each visit, is shown in Figure 10.5.

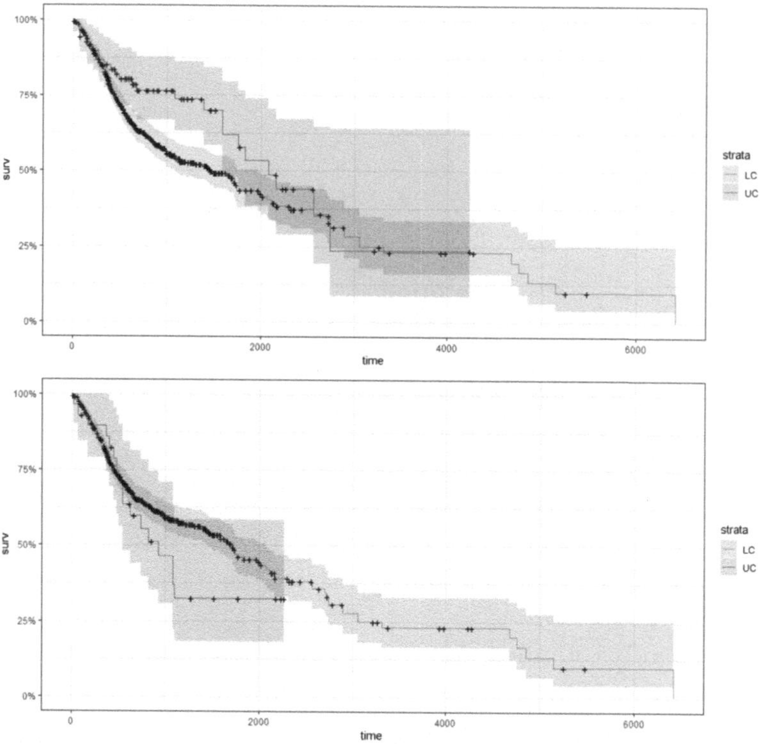

FIGURE 10.5: Kaplan Meier curves showing survival risk heterogeneity for different levels of expression values of biological markers.

A more detailed output of the coefficients and threshold values at each time point can be observed in the following commands. The optimal threshold values at each time point for each of the selected genes are shown in the output. We can find that the genes with positive coefficients at all time points are classified as Positive genes, whereas the genes with coefficient values consisting of both positive and negative values are classified as Volatile genes.

Due to the fitting of the CPH frailty model at each time point, we can find the unobserved subject-specific heterogeneity present at each time. The `random_part_sd` for each gene gives the corresponding frailty variances.

The `p_value` shows the minimum p-values at each time point to obtain the optimal thresholds. The `partition` explains the decile values (the p in equation 8) at which the gene expression provides the optimal threshold levels, and the `threshold` provides the optimal threshold value for the gene at that particular time-point.

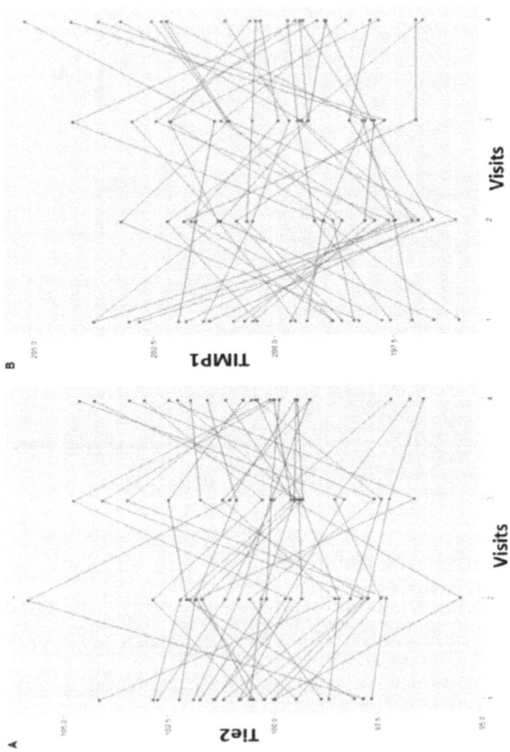

FIGURE 10.6: Longitudinal trajectory of 2 selected biomarkers in *mlclassCox* function.

10.8.1.6 mlclassKap

The `mlclassKap` works in a similar fashion with the learner model for selection is chosen as Kaplan Method. The result is shown in the following commands. Here we find that each of the Positive, Negative, and Volatile categories contain 2 genes. Some similarities are evident in the output of both the functions as `TLR4` is included in the Volatile Genes, and `IGFII` is categorized as Positive genes in both the functions. The detailed result for `mlclassKap` can be obtained similarly as previous.

10.9 Discussion

With the advancement of oncology treatments, the methodical analysis of genomic profiles carries significant importance. The researchers are thriving toward the more efficient formulation of mathematical and statistical models which can accurately forecast the disease trajectory. On one side, it is essential to gather reliable and sound data from the patients; on the other side, care should be taken to gather the relevant information from this large pool of data which will facilitate the necessary analysis to explore new insights into disease modeling. It will also lead to more focused research and optimized inference in the absence of redundant variables.

These aims can only get the picture when employed with computational power. Computational methods along with statistical algorithms can extract biological knowledge from the data. The knowledge is primarily in the form of relationships, which could be similarities about different gene structures, the relationship between gene expression and sample property, the effect of genomic profiles on the experimental conditions, etc. Machine learning portrays an important role in identifying these relationships. With the help of machine learning algorithms, the relationship can be quantified, and the genes with significant contributions can be identified.

In this study, we provided a comprehensive framework of survival analysis techniques along with classification and feature selection methods for gene expression data. The feature selection techniques explained in the article have been applied to high dimensional gene expression data to show their efficacy in selecting a few relevant markers.

It is pivotal to detect the nature of the effect of protein markers on overall survival. However, due to the data-driven approach, the chance of plausible randomness in the modeling cannot be ignored. Also, due to the employment of the cross-validation technique, the selected set of features is not unique at each draw.

So, it is advisable to perform the analysis a number of times to identify the most significant genes. We employed a classification technique using longitudinal values of gene expressions for the prediction of marker types and their functionalities. The classification algorithm is also capable of handling missing observations with the use of decision tree-based imputation

The heterogeneity in survival prediction is an important aspect of oncology research. Based on the selected features, we can perform frailty analysis to measure the randomness among the subjects. To accommodate the dependency between biological markers, we elaborated a threshold selection algorithm that will identify the risk heterogeneity at different expression levels using a minimum p-value.

Separate classes of expression levels are formed based on the optimal threshold levels, and we have checked the consistency in survival prediction for those classes. In this article, we have introduced an R package to perform the mentioned analytical procedures.

The **highMLR** package includes a comprehensive collection of survival analysis and heterogeneity analysis techniques for a high dimensional survival dataset. The main focus of this package is to deal with a high dimensional gene expression dataset obtained in oncology studies where it is required to select only a handful of essential features from a large pool of markers for further analysis.

Those markers are further analyzed to give a detailed impression of risk heterogeneity in expression values. The functions provided in this package give users the necessary options to better adjustment of models according to the dataset and the freedom to decide the prognostic measures required.

It is very rare to find a compact R package that provides a detailed analysis of heterogeneity identification and threshold value selection like the package **highMLR**. With the increasing interest in prognostic modeling and the development of treatment regimes, this package will fit the requirement of medical practitioners for a better understanding of disease progression.

Listing 10.6: Weibull distribution fitting in survival analysis using R

```
> test$Result
$result
      time  partition  threshold  min_           min_
                                  TIMP1_coef   TIMP1_p_value
70%      1    8          155.49    0.72           0.15
20%      2    3           84.82    0.54           0.47
60%      4    7          126.97   -1.05           0.10
20%1     5    3           85.93    0.71           0.38
      min_TIMP_time.1  partition.1  threshold.1  random_part_sd
70%   0.35             1            10             236.14
20%   0.10             2            9              191.69
60%   0.01             4            4              100.28
20%1  0.72             5            10             228.81
      min_IFNgammaR1   min_IFNgammaR1   min_IFNgammaR1
       _coef            _p_value        _random_part_sd
70%   1.13             0.09             0.40
20%   0.83             0.17             0.01
60%   -0.77            0.17             0.67
20%1  1.17             0.11             0.70
      time   partition   threshold  min_IGFII  min_IGFII
       .2      .2        .2         _coef      _p_value
70%     1      4         2156.07    1.00       0.11
20%     2      6         3545.33    1.71       0.00
60%     4      4         2385.33    1.73       0.09
20%1    5      6         3735.94    1.42       0.03
      min_IGFII          time.3  partition.3  threshold.3
      _random_part_sd
70%   0.01               1       8            2213.82
20%   0.01               2       8            2462.10
60%   0.01               4       6            1811.48
20%1  0.01               5       4            1340.23
      min_IFNgamma    min_IFNgamma   min_IFNgamma
       _coef           _p_value       _random_part_sd
70%    0.86           0.09            0.57
20%    1.08           0.03            0.40
60%    2.16           0.00            0.83
20%1   1.53           0.14            0.65
```

Listing 10.7: Weibull distribution fitting in survival analysis using R

```
> data(srdata)
> test=mlclassCox(m=50,n=69,idSurv=``OS",
                    idEvent=``event",Time=``Visit",
        s_ID=``ID",per=30,fold=3,data=srdata)
> test$GeneClassification
$Positive_Gene
     [,1]
[1,] ``IGFII"
[2,] ``IFNgamma"

$Negative_Gene
NULL

$Volatile_Gene
     [,1]
[1,] ``TIMP1"
[2,] ``IFNgammaR1"
[3,] ``Tie2"
[4,] ``TLR4"
```

Listing 10.8: Weibull distribution fitting in survival analysis using R

```
> test1=mlclassKap(m=50,n=69,idSurv=``OS",
                    idEvent=``event",Time=``Visit",
        s_ID=``ID",per=30,fold=3,data=srdata)
> test1$GeneClassification
$Positive_Gene
     [,1]
[1,] ``IGFII"
[2,] ``IFNbeta"

$Negative_Gene
     [,1]
[1,] ``TMEFF2"
[2,] ``TL1ATNFSF15"

$Volatile_Gene
     [,1]
[1,] ``TIMP3"
[2,] ``TLR4"
```

Bibliography

[1] David G Kleinbaum and Mitchel Klein. Introduction to logistic regression. In *Logistic regression*, pages 1–39. Springer, 2010.

[2] Ilari Kuitunen, Ville T Ponkilainen, Mikko M Uimonen, Antti Eskelinen, and Aleksi Reito. Testing the proportional hazards assumption in cox regression and dealing with possible non-proportionality in total joint arthroplasty research: methodological perspectives and review. *BMC Musculoskeletal Disorders*, 22(1):1–7, 2021.

[3] David R Cox. Regression models and life-tables. *Journal of the Royal Statistical Society: Series B (Methodological)*, 34(2):187–202, 1972.

[4] Odd Aalen. Nonparametric inference for a family of counting processes. *The Annals of Statistics*, pages 701–726, 1978.

[5] John P Klein. Modelling competing risks in cancer studies. *Statistics in Medicine*, 25(6):1015–1034, 2006.

[6] Boris Freidlin and Edward L Korn. Biomarker enrichment strategies: matching trial design to biomarker credentials. *Nature Reviews Clinical Oncology*, 11(2):81–90, 2014.

[7] Atanu Bhattacharjee, Gajendra Vishwakarma, and Abin Thomas. Detecting diagnostic accuracy of biomarkers through a bivariate beta distribution. *Turkiye Klinikleri Journal of Biostatistics*.

[8] Richard Simon and Aboubakar Maitournam. Evaluating the efficiency of targeted designs for randomized clinical trials. *Clinical Cancer Research*, 10(20):6759–6763, 2004.

[9] Vanita Noronha, Vijay Patil, Amit Joshi, Anuradha Chougule, Atanu Bhattacharjee, Rajiv Kumar, Sucheta More, Supriya Goud, Ashay

Karpe, Anant Ramaswamy, et al. Impact of exon 19 versus exon 21 egfr-activating mutation on outcomes with upfront pemetrexed–carboplatin chemotherapy. *ecancermedicalscience*, 11, 2017.

[10] Amit Joshi, Vanita Noronha, Vijay M Patil, Anuradha Chougule, Atanu Bhattacharjee, Rajiv Kumar, Supriya Goud, Sucheta More, Anant Ramaswamy, Ashay Karpe, et al. Efficacy of second-line pemetrexed-carboplatin in egfr-activating mutation-positive nsclc: Does exon 19 deletion differ from exon 21 mutation? *Chemotherapy Research and Practice*, 2017, 2017.

[11] Atanu Bhattacharjee and Vijay M Patil. Time-dependent area under the roc curve for optimum biological dose detection. *Turkiye Klinikleri Journal of Biostatistics*, 8(2), 2016.

[12] Ramesh Kumar, Jesna Jose, Gajendra K Vishwakarma, and Atanu Bhattacharjee. Alternative of clinical end point selection for metronomic studies. *IJS Short Reports*, 2(3):50–53, 2017.

[13] Amit Joshi, Vijay M Patil, Vanita Noronha, Joydeep Ghosh, Atanu Bhattacharjee, and Kumar Prabhash. A cross-sectional observation study regarding patients and their physician willingness to wait for driver mutation report in nonsmall-cell lung cancer. *Indian Journal of Medical and Paediatric Oncology: Official Journal of Indian Society of Medical & Paediatric Oncology*, 37(2):74, 2016.

[14] Vijay Maruti Patil, Vanita Noronha, Amit Joshi, Anuradha Bharat Choughule, Atanu Bhattacharjee, Rajiv Kumar, Supriya Goud, Sucheta More, Anant Ramaswamy, Ashay Karpe, et al. Phase iii study of gefitinib or pemetrexed with carboplatin in egfr-mutated advanced lung adenocarcinoma. *ESMO open*, 2(1):e000168, 2017.

[15] Vanita Noronha, Vijay M Patil, Amit Joshi, Atanu Bhattacharjee, Davinder Paul, Sachin Dhumal, Shashikant Juvekar, Supreeta Arya, and Kumar Prabhash. A tertiary care experience with paclitaxel and cetuximab as palliative chemotherapy in platinum sensitive and nonsensitive in head and neck cancers. *South Asian Journal of Cancer*, 6(1):11, 2017.

[16] Vijay M Patil, Vanita Noronha, Amit Joshi, Vamshi Muddu, Sachin Dhumal, Atanu Bhattacharjee, and Kumar Prabhash. Compliance with

neoadjuvant chemotherapy in t4 oral cancers: place, person, socioeconomic status, or assistance. *Journal of Global Oncology*, 1(2):65–72, 2015.

[17] Xian Zhou, Suyu Liu, Edward S Kim, Roy S Herbst, and J Jack Lee. Bayesian adaptive design for targeted therapy development in lung cancer-a step toward personalized medicine. *Clinical Trials*, 5(3):181–193, 2008.

[18] James H Albert and Siddhartha Chib. Bayesian analysis of binary and polychotomous response data. *Journal of the American statistical Association*, 88(422):669–679, 1993.

[19] Miranta Antoniou, Andrea L Jorgensen, and Ruwanthi Kolamunnage-Dona. Biomarker-guided adaptive trial designs in phase ii and phase iii: a methodological review. *PloS One*, 11(2):e0149803, 2016.

[20] Aiyi Liu, Chunling Liu, Qizhai Li, Kai F Yu, and Vivian W Yuan. A threshold sample-enrichment approach in a clinical trial with heterogeneous subpopulations. *Clinical Trials*, 7(5):537–545, 2010.

[21] Sue-Jane Wang, Robert T O'Neill, and HM James Hung. Approaches to evaluation of treatment effect in randomized clinical trials with genomic subset. *Pharmaceutical Statistics: The Journal of Applied Statistics in the Pharmaceutical Industry*, 6(3):227–244, 2007.

[22] Yosef Hochberg. A sharper bonferroni procedure for multiple tests of significance. *Biometrika*, 75(4):800–802, 1988.

[23] Maurice Berk, Cheryl Hemingway, Michael Levin, and Giovanni Montana. Longitudinal analysis of gene expression profiles using functional mixed-effects models. *Advanced statistical methods for the analysis of large data-sets*, pages 57–67, 2012.

[24] John D Storey, Wenzhong Xiao, Jeffrey T Leek, Ronald G Tompkins, and Ronald W Davis. Significance analysis of time course microarray experiments. *Proceedings of the National Academy of Sciences*, 102(36):12837–12842, 2005.

[25] Benoit Liquet, Kim-Anh Lê Cao, Hakim Hocini, and Rodolphe Thiébaut. A novel approach for biomarker selection and the integration

of repeated measures experiments from two assays. *BMC Bioinformatics*, 13(1):1–14, 2012.

[26] Lily Wang, Xi Chen, Russell D Wolfinger, Jeffrey L Franklin, Robert J Coffey, and Bing Zhang. A unified mixed effects model for gene set analysis of time course microarray experiments. *Statistical Applications in Genetics and Molecular Biology*, 8(1), 2009.

[27] Boris P Hejblum, Jason Skinner, and Rodolphe Thiébaut. Time-course gene set analysis for longitudinal gene expression data. *PLoS Computational Biology*, 11(6):e1004310, 2015.

[28] Agnieszka Król, Audrey Mauguen, Yassin Mazroui, Alexandre Laurent, Stefan Michiels, and Virginie Rondeau. Tutorial in joint modeling and prediction: a statistical software for correlated longitudinal outcomes, recurrent events and a terminal event. *arXiv preprint arXiv:1701.03675*, 2017.

[29] Robert M Elashoff, Gang Li, and Ning Li. A joint model for longitudinal measurements and survival data in the presence of multiple failure types. *Biometrics*, 64(3):762–771, 2008.

[30] Laura A Hatfield, Mark E Boye, Michelle D Hackshaw, and Bradley P Carlin. Multilevel bayesian models for survival times and longitudinal patient-reported outcomes with many zeros. *Journal of the American Statistical Association*, 107(499):875–885, 2012.

[31] Virginie Rondeau, Simone Mathoulin-Pelissier, Hélène Jacqmin-Gadda, Véronique Brouste, and Pierre Soubeyran. Joint frailty models for recurring events and death using maximum penalized likelihood estimation: application on cancer events. *Biostatistics*, 8(4):708–721, 2007.

[32] Lei Liu, Robert A Wolfe, and Xuelin Huang. Shared frailty models for recurrent events and a terminal event. *Biometrics*, 60(3):747–756, 2004.

[33] Yu Zhangsheng and Lei Liu. A joint model of recurrent events and a terminal event with a nonparametric covariate function. *Statistics in Medicine*, 30(22):2683–2695, 2011.

[34] Aurélien Belot, Virginie Rondeau, Laurent Remontet, Roch Giorgi, and CENSUR Working Survival Group. A joint frailty model to estimate

the recurrence process and the disease-specific mortality process without needing the cause of death. *Statistics in Medicine*, 33(18):3147–3166, 2014.

[35] Zhangsheng Yu, Lei Liu, Dawn M Bravata, and Linda S Williams. Joint model of recurrent events and a terminal event with time-varying coefficients. *Biometrical Journal*, 56(2):183–197, 2014.

[36] Yassin Mazroui, Audrey Mauguen, Simone Mathoulin-Pélissier, Gaetan MacGrogan, Véronique Brouste, and Virginie Rondeau. Time-varying coefficients in a multivariate frailty model: application to breast cancer recurrences of several types and death. *Lifetime Data Analysis*, 22(2):191–215, 2016.

[37] Debajyoti Sinha, Tapabrata Maiti, Joseph G Ibrahim, and Bichun Ouyang. Current methods for recurrent events data with dependent termination: a bayesian perspective. *Journal of the American Statistical Association*, 103(482):866–878, 2008.

[38] Achmad Efendi, Geert Molenberghs, Edmund Njeru Njagi, and Paul Dendale. A joint model for longitudinal continuous and time-to-event outcomes with direct marginal interpretation. *Biometrical Journal*, 55(4):572–588, 2013.

[39] Lei Liu and Xuelin Huang. Joint analysis of correlated repeated measures and recurrent events processes in the presence of death, with application to a study on acquired immune deficiency syndrome. *Journal of the Royal Statistical Society: Series C (Applied Statistics)*, 58(1):65–81, 2009.

[40] Sehee Kim, Donglin Zeng, Lloyd Chambless, and Yi Li. Joint models of longitudinal data and recurrent events with informative terminal event. *Statistics in Biosciences*, 4(2):262–281, 2012.

[41] Lei Liu, Xuelin Huang, and John O'Quigley. Analysis of longitudinal data in the presence of informative observational times and a dependent terminal event, with application to medical cost data. *Biometrics*, 64(3):950–958, 2008.

[42] Agnieszka Król, Loïc Ferrer, Jean-Pierre Pignon, Cécile Proust-Lima, Michel Ducreux, Olivier Bouché, Stefan Michiels, and Virginie Rondeau.

Joint model for left-censored longitudinal data, recurrent events and terminal event: predictive abilities of tumor burden for cancer evolution with application to the ffcd 2000–05 trial. *Biometrics*, 72(3):907–916, 2016.

[43] Dimitris Rizopoulos. Dynamic predictions and prospective accuracy in joint models for longitudinal and time-to-event data. *Biometrics*, 67(3):819–829, 2011.

[44] Cécile Proust-Lima and Jeremy MG Taylor. Development and validation of a dynamic prognostic tool for prostate cancer recurrence using repeated measures of posttreatment psa: a joint modeling approach. *Biostatistics*, 10(3):535–549, 2009.

[45] Audrey Mauguen, Bernard Rachet, Simone Mathoulin-Pélissier, Gaetan MacGrogan, Alexandre Laurent, and Virginie Rondeau. Dynamic prediction of risk of death using history of cancer recurrences in joint frailty models. *Statistics in Medicine*, 32(30):5366–5380, 2013.

[46] Paul Blanche, Cécile Proust-Lima, Lucie Loubère, Claudine Berr, Jean-François Dartigues, and Hélène Jacqmin-Gadda. Quantifying and comparing dynamic predictive accuracy of joint models for longitudinal marker and time-to-event in presence of censoring and competing risks. *Biometrics*, 71(1):102–113, 2015.

[47] Hans C Van Houwelingen and Hein Putter. Dynamic predicting by landmarking as an alternative for multi-state modeling: an application to acute lymphoid leukemia data. *Lifetime Data Analysis*, 14(4):447, 2008.

[48] Michael J Crowther, Richard D Riley, Jan A Staessen, Jiguang Wang, Francois Gueyffier, and Paul C Lambert. Individual patient data meta-analysis of survival data using poisson regression models. *BMC Medical Research Methodology*, 12(1):1–14, 2012.

[49] Andreas Wienke. *Frailty models in survival analysis.* Chapman and Hall/CRC, 2010.

[50] Luc Duchateau and Paul Janssen. *The frailty model.* Springer Science & Business Media, 2007.

[51] Philip Hougaard. Analysis of multivariate survival data, 2012.

[52] Virginie Rondeau, Yassin Mazroui, and Juan R Gonzalez. frailtypack: an r package for the analysis of correlated survival data with frailty models using penalized likelihood estimation or parametrical estimation. *J Stat Softw*, 47(4):1–28, 2012.

[53] Haiqun Lin and Daniel Zelterman. Modeling survival data: extending the cox model, 2002.

[54] Yvette Carter, Rebecca S Sippel, and Herbert Chen. Hypothyroidism after a cancer diagnosis: etiology, diagnosis, complications, and management. *The Oncologist*, 19(1):34, 2014.

[55] Allie Garcia-Serra, Robert J Amdur, Christopher G Morris, Ernest Mazzaferri, and William M Mendenhall. Thyroid function should be monitored following radiotherapy to the low neck. *American Journal of Clinical Oncology*, 28(3):255–258, 2005.

[56] Masashi Narazaki and Giovanna Tosato. Tumor cell populations differ in angiogenic activity: a model system for spontaneous angiogenic switch can tell us why. *Journal of the National Cancer Institute*, 98(5):294–295, 2006.

[57] Sara M Weis and David A Cheresh. Tumor angiogenesis: molecular pathways and therapeutic targets. *Nature Medicine*, 17(11):1359–1370, 2011.

[58] PKY Goon, GYH Lip, CJ Boos, PS Stonelake, and AD Blann. Circulating endothelial cells, endothelial progenitor cells, and endothelial microparticles in cancer. *Neoplasia*, 8(2):79–88, 2006.

[59] Jason L Rowand, Grace Martin, Gerald V Doyle, M Craig Miller, Michael S Pierce, Mark C Connelly, Chandra Rao, and Leon WMM Terstappen. Endothelial cells in peripheral blood of healthy subjects and patients with metastatic carcinomas. *Cytometry Part A: the Journal of the International Society for Analytical Cytology*, 71(2):105–113, 2007.

[60] Patrizia Mancuso, Alessandra Burlini, Giancarlo Pruneri, Aron Goldhirsch, Giovanni Martinelli, and Francesco Bertolini. Resting and activated endothelial cells are increased in the peripheral blood of cancer patients. *Blood, The Journal of the American Society of Hematology*, 97(11):3658–3661, 2001.

[61] LV Beerepoot, N Mehra, JSP Vermaat, BA Zonnenberg, MFGB Gebbink, and EE Voest. Increased levels of viable circulating endothelial cells are an indicator of progressive disease in cancer patients. *Annals of Oncology*, 15(1):139–145, 2004.

[62] Patrizia Mancuso, Marco Colleoni, Angelica Calleri, Laura Orlando, Patrick Maisonneuve, Giancarlo Pruneri, Alice Agliano, Aron Goldhirsch, Yuval Shaked, Robert S Kerbel, et al. Circulating endothelial-cell kinetics and viability predict survival in breast cancer patients receiving metronomic chemotherapy. *Blood*, 108(2):452–459, 2006.

[63] M Ilie, E Long, V Hofman, E Selva, C Bonnetaud, J Boyer, N Vénissac, C Sanfiorenzo, B Ferrua, CH Marquette, et al. Clinical value of circulating endothelial cells and of soluble cd146 levels in patients undergoing surgery for non-small cell lung cancer. *British Journal of Cancer*, 110(5):1236–1243, 2014.

[64] Atanu Bhattacharjee, Tapesh Bhattacharyya, Sanjana Santhosh, P Kiran, VP Sanudev Sadanandan, and M Geetha. Impact of waiting time for treatment on survival in patients undergoing radiotherapy for head and neck cancer. *Journal of Cancer Policy*, 13:1–4, 2017.

[65] Terry M Therneau and Thomas Lumley. Package 'survival'. *Survival analysis Published on CRAN*, 2:3, 2014.

[66] John Fox and Sanford Weisberg. *Cox proportional-hazards regression for survival data*. 2002.

[67] Thomas H Scheike and Mei-Jie Zhang. Analyzing competing risk data using the r timereg package. *Journal of Statistical Software*, 38(2), 2011.

[68] Torben Martinussen and Thomas H Scheike. Dynamic regression models for survival data. 2007.

[69] John M Kirkwood, Joseph G Ibrahim, Vernon K Sondak, Jon Richards, Lawrence E Flaherty, Marc S Ernstoff, Thomas J Smith, Uma Rao, Mary Steele, and Ronald H Blum. High-and low-dose interferon alfa-2b in high-risk melanoma: first analysis of intergroup trial e1690/s9111/c9190. *Journal of Clinical Oncology*, 18(12):2444–2458, 2000.

[70] John M Kirkwood, Joseph G Ibrahim, Jeffrey A Sosman, Vernon K Sondak, Sanjiv S Agarwala, Marc S Ernstoff, and Uma Rao. High-dose interferon alfa-2b significantly prolongs relapse-free and overall survival compared with the gm2-klh/qs-21 vaccine in patients with resected stage iib-iii melanoma: results of intergroup trial e1694/s9512/c509801. *Journal of Clinical Oncology*, 19(9):2370–2380, 2001.

[71] Y Chi and JG Ibrahim. A new class of joint models for longitudinal and survival data accomodating zero and non-zero cure fractions: A case study of an international breast cancer study group trial. *Stat Sin*, 17:445–462, 2007.

[72] Yueh-Yun Chi and Joseph G Ibrahim. Joint models for multivariate longitudinal and multivariate survival data. *Biometrics*, 62(2):432–445, 2006.

[73] Joseph W Hogan and Nan M Laird. Model-based approaches to analysing incomplete longitudinal and failure time data. *Statistics in Medicine*, 16(3):259–272, 1997.

[74] Mark D Schluchter. Methods for the analysis of informatively censored longitudinal data. *Statistics in Medicine*, 11(14-15):1861–1870, 1992.

[75] Atanu Bhattacharjee and Dilip C Nath. Joint longitudinal and survival data modelling: an application in anti-diabetes drug therapeutic effect. *Statistics*, 437, 2014.

[76] Anastasios A Tsiatis and Marie Davidian. Joint modeling of longitudinal and time-to-event data: an overview. *Statistica Sinica*, pages 809–834, 2004.

[77] Takeshi Emura, Wei-Chern Hsu, and Wen-Chi Chou. A survival tree based on stabilized score tests for high-dimensional covariates. *Journal of Applied Statistics*, pages 1–27, 2021.

[78] David G Beer, Sharon LR Kardia, Chiang-Ching Huang, Thomas J Giordano, Albert M Levin, David E Misek, Lin Lin, Guoan Chen, Tarek G Gharib, Dafydd G Thomas, et al. Gene-expression profiles predict survival of patients with lung adenocarcinoma. *Nature Medicine*, 8(8):816–824, 2002.

[79] Hsuan-Yu Chen, Sung-Liang Yu, Chun-Houh Chen, Gee-Chen Chang, Chih-Yi Chen, Ang Yuan, Chiou-Ling Cheng, Chien-Hsun Wang, Harn-Jing Terng, Shu-Fang Kao, et al. A five-gene signature and clinical outcome in non–small-cell lung cancer. *New England Journal of Medicine*, 356(1):11–20, 2007.

[80] Takeshi Emura, Shigeyuki Matsui, and Hsuan-Yu Chen. compound.cox: univariate feature selection and compound covariate for predicting survival. *Computer Methods and Programs in Biomedicine*, 168:21–37, 2019.

[81] Soyoung Kim and Ying Huang. Combining biomarkers for classification with covariate adjustment. *Statistics in medicine*, 36(15):2347–2362, 2017.

[82] Lance D Miller, Johanna Smeds, Joshy George, Vinsensius B Vega, Liza Vergara, Alexander Ploner, Yudi Pawitan, Per Hall, Sigrid Klaar, Edison T Liu, et al. An expression signature for p53 status in human breast cancer predicts mutation status, transcriptional effects, and patient survival. *Proceedings of the National Academy of Sciences*, 102(38):13550–13555, 2005.

[83] Christine Desmedt, Fanny Piette, Sherene Loi, Yixin Wang, Françoise Lallemand, Benjamin Haibe-Kains, Giuseppe Viale, Mauro Delorenzi, Yi Zhang, Mahasti Saghatchian d'Assignies, et al. Strong time dependence of the 76-gene prognostic signature for node-negative breast cancer patients in the transbig multicenter independent validation series. *Clinical cancer research*, 13(11):3207–3214, 2007.

[84] Kuo-Jang Kao, Kai-Ming Chang, Hui-Chi Hsu, and Andrew T Huang. Correlation of microarray-based breast cancer molecular subtypes and clinical outcomes: implications for treatment optimization. *BMC cancer*, 11(1):1–15, 2011.

[85] Sarah Dedeurwaerder, Christine Desmedt, Emilie Calonne, Sandeep K Singhal, Benjamin Haibe-Kains, Matthieu Defrance, Stefan Michiels, Michael Volkmar, Rachel Deplus, Judith Luciani, et al. Dna methylation profiling reveals a predominant immune component in breast cancers. *EMBO Molecular Medicine*, 3(12):726–741, 2011.

[86] Chi-Cheng Huang, Shih-Hsin Tu, Heng-Hui Lien, Jaan-Yeh Jeng, Ching-Shui Huang, Chi-Jung Huang, Liang-Chuan Lai, and Eric Y Chuang. Concurrent gene signatures for han chinese breast cancers. *PloS One*, 8(10):e76421, 2013.

[87] Ann Butler Nattinger, Mark S Gottlieb, Judith Veum, David Yahnke, and James S Goodwin. Geographic variation in the use of breast-conserving treatment for breast cancer. *New England Journal of Medicine*, 326(17):1102–1107, 1992.

[88] Roger J Lewis. An introduction to classification and regression tree (cart) analysis. In *Annual meeting of the society for academic emergency medicine in San Francisco, California*, volume 14. Citeseer, 2000.

[89] M Luz Calle, Guillermo Hough, Ana Curia, and Guadalupe Gómez. Bayesian survival analysis modeling applied to sensory shelf life of foods. *Food Quality and Preference*, 17(3-4):307–312, 2006.

[90] Farid E Ahmed, Paul W Vos, and Don Holbert. Modeling survival in colon cancer: a methodological review. *Molecular cancer*, 6(1):1–12, 2007.

[91] Imran Kurt Omurlu, Kazim Ozdamar, and Mevlut Ture. Comparison of bayesian survival analysis and cox regression analysis in simulated and breast cancer data sets. *Expert Systems with Applications*, 36(8):11341–11346, 2009.

[92] MCM Wong, KF Lam, and ECM Lo. Bayesian analysis of clustered interval-censored data. *Journal of Dental Research*, 84(9):817–821, 2005.

[93] Atanu Bhattacharjee. Application of bayesian approach in cancer clinical trial. *World Journal of Oncology*, 5(3):109, 2014.

[94] Dennis V Lindley. The bayesian approach to statistics. Technical report, California Univ Berkeley Operations Research Center, 1980.

[95] Torsten Hothorn and Peter Bühlmann. Model-based boosting in high dimensions. *Bioinformatics*, 22(22):2828–2829, 2006.

[96] Peter Bühlmann and Bin Yu. Boosting with the l 2 loss: regression and classification. *Journal of the American Statistical Association*, 98(462):324–339, 2003.

[97] Peter Buehlmann. Boosting for high-dimensional linear models. *The Annals of Statistics*, 34(2):559–583, 2006.

[98] Peter Bühlmann and Torsten Hothorn. Boosting algorithms: Regularization, prediction and model fitting. *Statistical Science*, 22(4):477–505, 2007.

[99] Avrim L Blum and Pat Langley. Selection of relevant features and examples in machine learning. *Artificial intelligence*, 97(1-2):245–271, 1997.

[100] Robert N Rodriguez. Statistical model building for large, complex data: Five new directions in sas/stat software. In *Proceedings of the SAS Global Forum 2016 Conference. Cary, NC: SAS Institute Inc. http://support. sas. com/resources/papers/proceedings16/SAS4900-2016. pdf*, 2016.

[101] Douglas M Hawkins. The problem of overfitting. *Journal of Chemical Information and Computer Sciences*, 44(1):1–12, 2004.

[102] Christopher J Greenwood, George J Youssef, Primrose Letcher, Jacqui A Macdonald, Lauryn J Hagg, Ann Sanson, Jenn Mcintosh, Delyse M Hutchinson, John W Toumbourou, Matthew Fuller-Tyszkiewicz, et al. A comparison of penalised regression methods for informing the selection of predictive markers. *Plos One*, 15(11):e0242730, 2020.

[103] Robert Tibshirani. Regression shrinkage and selection via the lasso. *Journal of the Royal Statistical Society: Series B (Methodological)*, 58(1):267–288, 1996.

[104] Hui Zou and Trevor Hastie. Regularization and variable selection via the elastic net. *Journal of the Royal Statistical Society: Series B (Statistical Methodology)*, 67(2):301–320, 2005.

[105] Bradley Efron, Trevor Hastie, Iain Johnstone, and Robert Tibshirani. Least angle regression. *The Annals of Statistics*, 32(2):407–499, 2004.

[106] Shuangge Ma and Jian Huang. Additive risk survival model with microarray data. *BMC Bioinformatics*, 8(1):1–10, 2007.

[107] Zhifu Sun, Ping Yang, Marie-Christine Aubry, Farhad Kosari, Chiaki Endo, Julian Molina, and George Vasmatzis. Can gene expression profiling predict survival for patients with squamous cell carcinoma of the lung? *Molecular Cancer*, 3(1):1–10, 2004.

[108] Dennis A Wigle, Igor Jurisica, Niki Radulovich, Melania Pintilie, Janet Rossant, Ni Liu, Chao Lu, James Woodgett, Isolde Seiden, Michael Johnston, et al. Molecular profiling of non-small cell lung cancer and correlation with disease-free survival. *Cancer Research*, 62(11):3005–3008, 2002.

[109] Alain C Borczuk, Lori Shah, Gregory DN Pearson, Kristin L Walter, Liqun Wang, John HM Austin, Richard A Friedman, and Charles A Powell. Molecular signatures in biopsy specimens of lung cancer. *American Journal of Respiratory and Critical Care Medicine*, 170(2):167–174, 2004.

[110] David Engler and Yi Li. Survival analysis with high-dimensional covariates: an application in microarray studies. *Statistical Applications in Genetics and Molecular Biology*, 8(1), 2009.

[111] Jianqing Fan and Jinchi Lv. Sure independence screening for ultrahigh dimensional feature space. *Journal of the Royal Statistical Society: Series B (Statistical Methodology)*, 70(5):849–911, 2008.

[112] Jianqing Fan and Jinchi Lv. A selective overview of variable selection in high dimensional feature space. *Statistica Sinica*, 20(1):101, 2010.

[113] Sandrine Dudoit, Juliet Popper Shaffer, and Jennifer C Boldrick. Multiple hypothesis testing in microarray experiments. *Statistical Science*, 18(1):71–103, 2003.

[114] John D Storey and Robert Tibshirani. Statistical significance for genomewide studies. *Proceedings of the National Academy of Sciences*, 100(16):9440–9445, 2003.

[115] Jianqing Fan and Yi Ren. Statistical analysis of dna microarray data in cancer research. *Clinical Cancer Research*, 12(15):4469–4473, 2006.

[116] Bradley Efron. Correlation and large-scale simultaneous significance testing. *Journal of the American Statistical Association*, 102(477):93–103, 2007.

[117] Jianqing Fan, Richard Samworth, and Yichao Wu. Ultrahigh dimensional variable selection: beyond the linear model. *arXiv preprint arXiv:0812.3201*, 2008.

[118] Hongzhe Li and Yihui Luan. Kernel cox regression models for linking gene expression profiles to censored survival data. In *Biocomputing 2003*, pages 65–76. World Scientific, 2002.

[119] Jiang Gui and Hongzhe Li. Penalized cox regression analysis in the high-dimensional and low-sample size settings, with applications to microarray gene expression data. *Bioinformatics*, 21(13):3001–3008, 2005.

[120] Mark R Segal. Microarray gene expression data with linked survival phenotypes: diffuse large-b-cell lymphoma revisited. *Biostatistics*, 7(2):268–285, 2006.

[121] Mee Young Park and Trevor Hastie. L1-regularization path algorithm for generalized linear models. *Journal of the Royal Statistical Society: Series B (Statistical Methodology)*, 69(4):659–677, 2007.

[122] Naijun Sha, Mahlet G Tadesse, and Marina Vannucci. Bayesian variable selection for the analysis of microarray data with censored outcomes. *Bioinformatics*, 22(18):2262–2268, 2006.

[123] G Schwarz. Estimating thedimension of a model. annals ofstatistics, 6 (2), 461–464, 1978.

[124] Hirotogu Akaike. Information theory and an extension of the maximum likelihood principle. In *Selected papers of hirotugu akaike*, pages 199–213. Springer, 1998.

[125] Robert E Kass and Larry Wasserman. A reference bayesian test for nested hypotheses and its relationship to the schwarz criterion. *Journal of the American Statistical Association*, 90(431):928–934, 1995.

[126] James W Vaupel, Kenneth G Manton, and Eric Stallard. The impact of heterogeneity in individual frailty on the dynamics of mortality. *Demography*, 16(3):439–454, 1979.

[127] Major Greenwood and G Udny Yule. An inquiry into the nature of frequency distributions representative of multiple happenings with particular reference to the occurrence of multiple attacks of disease or of repeated accidents. *Journal of the Royal Statistical Society*, 83(2):255–279, 1920.

[128] David G Clayton. A model for association in bivariate life tables and its application in epidemiological studies of familial tendency in chronic disease incidence. *Biometrika*, 65(1):141–151, 1978.

[129] John P Klein. Semiparametric estimation of random effects using the cox model based on the em algorithm. *Biometrics*, pages 795–806, 1992.

[130] LJ Wei, and David V Glidden. An overview of statistical methods for multiple failure time data in clinical trials. *Statistics in Medicine*, 16(8):833–839, 1997.

[131] Philip Hougaard. Frailty models for survival data. *Lifetime Data Analysis*, 1(3):255–273, 1995.

[132] Olga Theou, Mario Ulises Pérez-Zepeda, Alexandra M van der Valk, Samuel D Searle, Susan E Howlett, and Kenneth Rockwood. A classification tree to assist with routine scoring of the clinical frailty scale. *Age and ageing*, 50(4):1406–1411, 2021.

[133] R Core Team et al. R: A language and environment for statistical computing. 2021.

[134] TA Balan and H Putter. frailtyem: fitting frailty models with the em algorithm. *R package version 0.5*, 4, 2017.

[135] Il Do Ha, Maengseok Noh, and Youngjo Lee. frailtyhl: A package for fitting frailty models with h-likelihood. *The R Journal*, 4(2):28–36, 2012.

[136] John V Monaco, Malka Gorfine, and Li Hsu. General semiparametric shared frailty model: Estimation and simulation with frailtysurv. *Journal of Statistical Software*, 86, 2018.

[137] Marco Munda, Federico Rotolo, Catherine Legrand, et al. Parfm: parametric frailty models in r. *Journal of Statistical Software*, 51(11):1–20, 2012.

[138] Jeremy Walston, Thomas N Robinson, Susan Zieman, Frances McFarland, Christopher R Carpenter, Keri N Althoff, Melissa K Andrew, Caroline S Blaum, Patrick J Brown, Brian Buta, et al. Integrating frailty research into the medical specialties—report from a u13 conference. *Journal of the American Geriatrics Society*, 65(10):2134–2139, 2017.

[139] Robert M Califf. Biomarker definitions and their applications. *Experimental Biology and Medicine*, 243(3):213–221, 2018.

[140] Atanu Bhattacharjee, Jacinth Rajendra, Rajesh Dikshit, and Shilpee Dutt. Her2 borderline is a negative prognostic factor for primary malignant breast cancer. *Breast Cancer Research and Treatment*, 181(1):225–231, 2020.

[141] Hein Putter, Marta Fiocco, and Ronald B Geskus. Tutorial in biostatistics: competing risks and multi-state models. *Statistics in Medicine*, 26(11):2389–2430, 2007.

[142] Shigeyuki Matsui. Genomic biomarkers for personalized medicine: development and validation in clinical studies. *Computational and Mathematical Methods in Medicine*, 2013, 2013.

[143] Jung-Yun Lee, Hee Seung Kim, Dong Hoon Suh, Mi-Kyung Kim, Hyun Hoon Chung, and Yong-Sang Song. Ovarian cancer biomarker discovery based on genomic approaches. *Journal of Cancer Prevention*, 18(4):298, 2013.

[144] Emilio Marengo and Elisa Robotti. Biomarkers for pancreatic cancer: recent achievements in proteomics and genomics through classical and multivariate statistical methods. *World Journal of Gastroenterology: WJG*, 20(37):13325, 2014.

[145] André Ahr, Thomas Karn, Christine Solbach, Tanja Seiter, Klaus Strebhardt, Uwe Holtrich, and Manfred Kaufmann. Identification of high risk breast-cancer patients by gene expression profiling. *The Lancet*, 359(9301):131–132, 2002.

[146] William D Shannon, Mark A Watson, Arie Perry, and Keith Rich. Mantel statistics to correlate gene expression levels from microarrays with clinical covariates. *Genetic Epidemiology*, 23(1):87–96, 2002.

[147] Kaushik Ghosh and Ram C Tiwari. Prediction of us cancer mortality counts using semiparametric bayesian techniques. *Journal of the American Statistical Association*, 102(477):7–15, 2007.

[148] Charles E Antoniak. Mixtures of dirichlet processes with applications to bayesian nonparametric problems. *The Annals of Statistics*, pages 1152–1174, 1974.

[149] Jon T Schnute. A general framework for developing sequential fisheries models. *Canadian Journal of Fisheries and Aquatic Sciences*, 51(8):1676–1688, 1994.

[150] Richard J Meinhold and Nozer D Singpurwalla. Understanding the kalman filter. *The American Statistician*, 37(2):123–127, 1983.

[151] Bradley P Carlin, Nicholas G Polson, and David S Stoffer. A monte carlo approach to nonnormal and nonlinear state-space modeling. *Journal of the American Statistical Association*, 87(418):493–500, 1992.

[152] P Jeffrey Harrison and Colin F Stevens. Bayesian forecasting. *Journal of the Royal Statistical Society: Series B (Methodological)*, 38(3):205–228, 1976.

[153] Kenneth H Reckhow. Bayesian inference in non-replicated ecological studies. *Ecology*, 71(6):2053–2059, 1990.

[154] Paul R Wade. Bayesian methods in conservation biology. *Conservation Biology*, 14(5):1308–1316, 2000.

[155] Chantal Guihenneuc-Jouyaux, Sylvia Richardson, and Ira M Longini Jr. Modeling markers of disease progression by a hidden markov process: application to characterizing cd4 cell decline. *Biometrics*, 56(3):733–741, 2000.

[156] Philip D O'Neill. A tutorial introduction to bayesian inference for stochastic epidemic models using markov chain monte carlo methods. *Mathematical Biosciences*, 180(1-2):103–114, 2002.

[157] Tanja Kostic, Patrick Butaye, and Jacques Schrenzel. *Detection of Highly Dangerous Pathogens: Microarray Methods for BSL 3 and BSL 4 Agents.* John Wiley & Sons, 2009.

[158] Hua Liao, Xiaozhen J Liu, Jonathan L Blank, David C Bouck, Hugues Bernard, Khristofer Garcia, and Eric S Lightcap. Quantitative proteomic analysis of cellular protein modulation upon inhibition of the nedd8-activating enzyme by mln4924. *Molecular & Cellular Proteomics*, 10(11), 2011.

[159] Bernard L Welch. The generalization of 'student's problem when several different population varlances are involved. *Biometrika*, 34(1-2):28–35, 1947.

[160] John Geweke and Hisashi Tanizaki. Bayesian estimation of state-space models using the metropolis–hastings algorithm within gibbs sampling. *Computational Statistics & Data Analysis*, 37(2):151–170, 2001.

[161] Ramon C Littell, Jane Pendergast, and Ranjini Natarajan. Tutorial in biostatistics modelling covariance structure in the analysis of repeated measures data. *Tutorials in Biostatistics, Tutorials in Biostatistics: Statistical Modelling of Complex Medical Data*, 2:159, 2005.

[162] Donald Hedeker and Robert D Gibbons. *Longitudinal data analysis.* Wiley-Interscience, 2006.

[163] Otília Menyhart, Boglárka Weltz, and Balázs Győrffy. Multipletesting. com: a tool for life science researchers for multiple hypothesis testing correction. *PloS one*, 16(6):e0245824, 2021.

[164] Martin A Tanner and Wing Hung Wong. The calculation of posterior distributions by data augmentation. *Journal of the American statistical Association*, 82(398):528–540, 1987.

[165] Peter Diggle, Peter J Diggle, Patrick Heagerty, Kung-Yee Liang, Scott Zeger, et al. *Analysis of longitudinal data.* Oxford university press, 2002.

[166] Yu Chuan Tai and Terence P Speed. Statistical analysis of microarray time course data. In *DNA Microarrays*, pages 257–280. Bios Scientific Publishers Limited, Oxford, 2005.

[167] James M Tepperman, Matthew E Hudson, Rajnish Khanna, Tong Zhu, Sherman H Chang, Xun Wang, and Peter H Quail. Expression profiling of phyb mutant demonstrates substantial contribution of other phytochromes to red-light-regulated gene expression during seedling deetiolation. *The Plant Journal*, 38(5):725–739, 2004.

[168] Juha-Pekka Himanen, Michael J Chumley, Martin Lackmann, Chen Li, William A Barton, Phillip D Jeffrey, Christopher Vearing, Detlef Geleick, David A Feldheim, Andrew W Boyd, et al. Repelling class discrimination: ephrin-a5 binds to and activates ephb2 receptor signaling. *Nature Neuroscience*, 7(5):501–509, 2004.

[169] Wei Qi, Andrew Ratanatharathorn, Martin Gevonden, Richard Bryant, Douglas Delahanty, Yutaka Matsuoka, Miranda Olff, Terri deRoon Cassini, Ulrich Schnyder, Soraya Seedat, et al. Application of data pooling to longitudinal studies of early post-traumatic stress disorder (ptsd): the international consortium to predict ptsd (icpp) project. *European Journal of Psychotraumatology*, 9(1):1476442, 2018.

[170] Pavel Tomancak, Amy Beaton, Richard Weiszmann, Elaine Kwan, ShengQiang Shu, Suzanna E Lewis, Stephen Richards, Michael Ashburner, Volker Hartenstein, Susan E Celniker, et al. Systematic determination of patterns of gene expression during drosophila embryogenesis. *Genome Biology*, 3(12):1–14, 2002.

[171] Szu-Peng Yang and Takeshi Emura. A bayesian approach with generalized ridge estimation for high-dimensional regression and testing. *Communications in Statistics-Simulation and Computation*, 46(8):6083–6105, 2017.

[172] Nikita Singh, Mukesh Kumar, Atanu Bhattacharjee, Prashant Sonker, and Agni Saroj. Identification of novel biomarkers and enriched pathways involved in lung cancer using statistical and bioinformatics techniques. *Authorea Preprints*, 2020.

[173] Nimisha Tripathi, Sneha Keshari, Pallavi Shahi, Poonam Maurya, Atanu Bhattacharjee, Kushal Gupta, Sanjay Talole, and Mukesh Kumar. Human papillomavirus elevated genetic biomarker signature by statistical algorithm. *Journal of Cellular Physiology*, 235(12):9922–9932, 2020.

[174] Bradley Efron, Robert Tibshirani, John D Storey, and Virginia Tusher. Empirical bayes analysis of a microarray experiment. *Journal of the American Statistical Association*, 96(456):1151–1160, 2001.

[175] Andrew O Finley, Sudipto Banerjee, and Bradley P Carlin. spbayes: an r package for univariate and multivariate hierarchical point-referenced spatial models. *Journal of Statistical Software*, 19(4):1, 2007.

[176] Sebastian Meltzer, Erta Kalanxhi, Helga Helseth Hektoen, Svein Dueland, Kjersti Flatmark, Kathrine Røe Redalen, and Anne Hansen Ree. Systemic release of osteoprotegerin during oxaliplatin-containing induction chemotherapy and favorable systemic outcome of sequential radiotherapy in rectal cancer. *Oncotarget*, 7(23):34907, 2016.

[177] Reinhard Marks, Sylvia Kock, Claudia Wehr, Jürgen Finke, Martin Werner, Justus Duyster, and Paul Fisch. High throughput t cell receptor sequencing augments diagnosis and response assessment in patients with angioimmunoblastic t cell lymphoma. *Blood*, 130(Supplement 1):1526–1526, 2017.

[178] Eddy C Hsueh, James R DeBloom, Jonathan Lee, Jeffrey J Sussman, Kyle R Covington, Brooke Middlebrook, Clare Johnson, Robert W Cook, Craig L Slingluff, and Kelly M McMasters. Interim analysis of survival in a prospective, multi-center registry cohort of cutaneous melanoma tested with a prognostic 31-gene expression profile test. *Journal of Hematology & Oncology*, 10(1):1–8, 2017.

[179] David R Cox. Regression models and life-tables. breakthroughs in statistics. *New York*, 372:527–41, 1992.

[180] Robert A McLean, William L Sanders, and Walter W Stroup. A unified approach to mixed linear models. *The American Statistician*, 45(1):54–64, 1991.

[181] Dimitris Rizopoulos. Joint models for longitudinal and time-to-event data: With applications in r. 2012.

[182] Maria Sudell, Ruwanthi Kolamunnage-Dona, and Catrin Tudur-Smith. Joint models for longitudinal and time-to-event data: a review of reporting quality with a view to meta-analysis. *BMC Medical Research Methodology*, 16(1):1–11, 2016.

[183] Dimitris Rizopoulos. The r package jmbayes for fitting joint models for longitudinal and time-to-event data using mcmc. *arXiv preprint arXiv:1404.7625*, 2014.

[184] Graeme L Hickey, Pete Philipson, Andrea Jorgensen, and Ruwanthi Kolamunnage-Dona. Joint models of longitudinal and time-to-event data with more than one event time outcome: a review. *The International Journal of Biostatistics*, 14(1), 2018.

[185] Robin Henderson, Peter Diggle, and Angela Dobson. Joint modelling of longitudinal measurements and event time data. *Biostatistics*, 1(4):465–480, 2000.

[186] Yangxin Huang, Lang Wu, Grace Y Yi, and Wenbin Lu. Joint models and their applications. 2012.

[187] Donna K Pauler and Dianne M Finkelstein. Predicting time to prostate cancer recurrence based on joint models for non-linear longitudinal biomarkers and event time outcomes. *Statistics in Medicine*, 21(24):3897–3911, 2002.

[188] Joseph W Hogan and Nan M Laird. Increasing efficiency from censored survival data by using random effects to model longitudinal covariates. *Statistical Methods in Medical Research*, 7(1):28–48, 1998.

[189] Alison Yan Zhang, Karen Chiam, Ygal Haupt, Stephen B Fox, Simone Birch, Wayne Tilley, Lisa Butler, Karen E Knudsen, Christopher Cornstock, Krishan Rasiah, et al. An analysis of multiple biomarkers to better predict prostate cancer metastasis and death after radical prostatectomy. 2018.

[190] Mario Stampanoni Bassi, Ennio Iezzi, Girolama A Marfia, Ilaria Simonelli, Alessandra Musella, Georgia Mandolesi, Diego Fresegna, Patrizio Pasqualetti, Roberto Furlan, Annamaria Finardi, et al. Platelet-derived growth factor predicts prolonged relapse-free period in multiple sclerosis. *Journal of Neuroinflammation*, 15(1):1–9, 2018.

[191] David Roxbee Cox and David Oakes. *Analysis of survival data*. Chapman and Hall/CRC, 2018.

[192] John D Kalbfleisch and Ross L Prentice. *The statistical analysis of failure time data*. John Wiley & Sons, 2011.

[193] John P Klein and Melvin L Moeschberger. *Survival analysis: techniques for censored and truncated data*, volume 2. Springer, 2003.

[194] Terry M Therneau. Extending the cox model. In *Proceedings of the First Seattle symposium in biostatistics*, pages 51–84. Springer, 1997.

[195] Judith D Singer, John B Willett, John B Willett, et al. *Applied longitudinal data analysis: Modeling change and event occurrence*. Oxford university press, 2003.

[196] Sophia Rabe-Hesketh and Anders Skrondal. *Multilevel and longitudinal modeling using Stata*. STATA press, 2008.

[197] Benoit Liquet and Jérémie Riou. Cpmcglm: an r package for p-value adjustment when looking for an optimal transformation of a single explanatory variable in generalized linear models. *BMC Medical Research Methodology*, 19(1):1–8, 2019.

[198] Reza Hashemi and Daniel Commenges. Correction of the p-value after multiple tests in a cox proportional hazard model. *Lifetime Data Analysis*, 8(4):335–348, 2002.

[199] Glenn W Brier et al. Verification of forecasts expressed in terms of probability. *Monthly weather review*, 78(1):1–3, 1950.

[200] Geert Verbeke, Steffen Fieuws, Geert Molenberghs, and Marie Davidian. The analysis of multivariate longitudinal data: a review. *Statistical Methods in Medical Research*, 23(1):42–59, 2014.

[201] Graeme L Hickey, Pete Philipson, Andrea Jorgensen, and Ruwanthi Kolamunnage-Dona. Joint modelling of time-to-event and multivariate longitudinal outcomes: recent developments and issues. *BMC Medical Research Methodology*, 16(1):1–15, 2016.

[202] Cheryl L Faucett and Duncan C Thomas. Simultaneously modelling censored survival data and repeatedly measured covariates: a gibbs sampling approach. *Statistics in Medicine*, 15(15):1663–1685, 1996.

[203] Cheryl L Faucett, Nathaniel Schenker, and Robert M Elashoff. Analysis of censored survival data with intermittently observed time-dependent binary covariates. *Journal of the American Statistical Association*, 93(442):427–437, 1998.

[204] Yan Wang and Jeremy M G Taylor. Jointly modeling longitudinal and event time data with application to acquired immunodeficiency syndrome. *Journal of the American Statistical Association*, 96(455):895–905, 2001.

[205] Jane Xu and Scott L Zeger. Joint analysis of longitudinal data comprising repeated measures and times to events. *Journal of the Royal Statistical Society: Series C (Applied Statistics)*, 50(3):375–387, 2001.

[206] Elizabeth R. Brown and Joseph G. Ibrahim. A bayesian semiparametric joint hierarchical model for longitudinal and survival data. *Biometrics*, 59(2):221–228, 2003.

[207] Yueh-Yun Chi and Joseph G Ibrahim. Bayesian approaches to joint longitudinal and survival models accommodating both zero and nonzero cure fractions. *Statistica Sinica*, pages 445–462, 2007.

[208] Dimitris Rizopoulos and Emmanuel Lesaffre. Introduction to the special issue on joint modelling techniques. *Statistical methods in medical research*, 23(1):3–10, 2014.

[209] Alexander C McLain and Sujit K Ghosh. Efficient sieve maximum likelihood estimation of time-transformation models. *Journal of Statistical Theory and Practice*, 7(2):285–303, 2013.

[210] Douglas Bates, Martin Mächler, Ben Bolker, and Steve Walker. Fitting linear mixed-effects models using lme4. *arXiv preprint arXiv:1406.5823*, 2014.

[211] Ngayee J Law, Jeremy MG Taylor, and Howard Sandler. The joint modeling of a longitudinal disease progression marker and the failure time process in the presence of cure. *Biostatistics*, 3(4):547–563, 2002.

[212] Julie Marin and S Blair Hedges. Undersampling genomes has biased time and rate estimates throughout the tree of life. *Molecular Biology and Evolution*, 35(8):2077–2084, 2018.

[213] Mickaël Canouil, Beverley Balkau, Ronan Roussel, Philippe Froguel, and Ghislain Rocheleau. Jointly modelling single nucleotide polymorphisms with longitudinal and time-to-event trait: An application to type 2 diabetes and fasting plasma glucose. *Frontiers in Genetics*, 9:210, 2018.

[214] Atanu Bhattacharjee, Gajendra K Vishwakarma, and Abin Thomas. Bayesian state-space modeling in gene expression data analysis: An application with biomarker prediction. *Mathematical Biosciences*, 305:96–101, 2018.

[215] L. Yu and H. Liu. Feature selection for high-dimensional data: A fast correlation-based filter solution. *The 20th International Conference on Machine Learning (ICML-03)*, pages 856–863, 2003.

[216] V.B. Canedo, N.S. Maroño, and A.A. Betanzos. A review of feature selection methods on synthetic data. *Knowl. Inf. Syst.*, 34:483–519, 2013.

[217] J. Li, K. Cheng, S. Wang, F. Morstatter, R.P. Trevino, J. Tang, and H. Liu. Feature Selection: A Data Perspective. *ACM Comput. Surv.*, 50(6):1–45, 2017.

[218] T. Li, C. Zhang, and M. Ogihara. A comparative study of feature selection and multiclass classification methods for tissue classification based on gene expression. *Bioinformatics*, 20(15):2429–2437, 2004.

[219] C. Ambroise and G.J. McLachlan. Selection bias in gene extraction on the basis of microarray gene-expression data. *Proc. Nat. Acad. Sci.*, 99(10):6562–6566, 2002.

[220] C. Ding and H. Peng. Minimum redundancy feature selection from microarray gene expression data. *J. Bioinform. Comput. Biol.*, 3(2):185–205, 2005.

[221] J.J. Heckman. Sample Selection Bias as a Specification Error. *Econometrica*, 47(1):153–161, 1979.

[222] A. Jović, K. Brkić, and N. Bogunović. A review of feature selection methods with applications. *38th International Convention on Information and Communication Technology, Electronics and Microelectronics (MIPRO), IEEE*, page 1200–1205, 2015.

[223] N.S. Maroño, A.A. Betanzos, and M.T. Sanromán. Filter methods for feature selection–a comparative study. *International Conference on Intelligent Data Engineering and Automated Learning*, Springer, pages 178–187, 2007.

[224] I. Guyon and A. Elisseeff. An Introduction to Variable and Feature Selection. *J. Mach. Learn. Res.*, 3:1157–1182, 2003.

[225] T.N. Lal, O. Chapelle, J. Weston, and A. Elisseeff. Embedded methods. *Feature Extraction*, Springer, pages 137–165, 2006.

[226] J. Friedman, T. Hastie, and R. Tibshirani. Regularization paths for generalized linear models via coordinate descent. *J. Stat. Softw.*, 33(1):1–22, 2010.

[227] E.P. Xing, M.I. Jordan, and R.M. Karp. Feature selection for high-dimensional genomic microarray data. *ICML*, 1:601–608, 2001.

[228] G. Chandrashekar and F. Sahin. A survey on feature selection methods. *Comp. Elec. Eng.*, 40:16–28, 2014.

[229] B.C. Kuo, H.H. Ho, C.H. Li, C.C. Hung, and J.S. Taur. A kernel-based feature selection method for SVM with RBF kernel for hyperspectral image classification. *IEEE J. Sel. Top. Appl. Earth Obs. Remote Sens.*, 7(1):317–326, 2013.

[230] Z. Li, W. Xie, and T. Liu. Efficient feature selection and classification for microarray data. *PLoS One*, 13(8), 2018.

[231] V.B. Canedo, N.S. Marono, A.A. Betanzos, J.M. Benítez, and F. Herrera. A review of microarray datasets and applied feature selection methods. *Inform. Sci.*, 282:111–135, 2014.

[232] S. Ma and J. Huang. Penalized feature selection and classification in bioinformatics. *Brief. Bioinform.*, 9(5):392–403, 2008.

[233] G.K. Vishwakarma, A. Bhattacharjee, and A. Thomas. A weight function method for selection of proteins to predict an outcome using protein expression data. *J. Comput. Appl. Math.*, 391:113465, 2021.

[234] J.E. Cavanaugh. Unifying the Derivations for the Akaike and Corrected Akaike Information Criteria. *Stat. Prob. Letters*, 33:201–208, 1997.

[235] J. Friedman, T. Hastie, and R. Tibshirani. glmnet: Lasso and Elastic-Net Regularized Generalized Linear Models. *R Package Version*, 1(4), 2009.

[236] H. Zou and T. Hastie. Regularization and variable selection via the elastic net. *J. R. Stat. Soc. Ser. B Stat. Methodol.*, 67(2):301–320, 2005.

[237] R. Tibshirani. Regression shrinkage and selection via the lasso. *J. R. Stat. Soc. Ser. B Stat. Methodol*, 58(1):267–288, 1996.

[238] S.P. Yang and T. Emura. A bayesian approach with generalized ridge estimation for high-dimensional regression and testing. *Comm. Statist. Simu. Comput.*, 46(8):6083–6105, 2017.

[239] J. Tang, S. Alelyani, and H. Liu. *Feature selection for classification: A review.* 2014.

[240] S. Nogueira. Quantifying the stability of feature selection. *Thesis, Univ. Manc.*, 2018.

[241] L. Jiang, N. Haiminen, A.P. Carrieri, S. Huang, Y. Vázquez-Baeza, L. Parida, H.C. Kim, A.D. Swafford, R. Knight, and L. Natarajan. Utilizing stability criteria in choosing feature selection methods yields reproducible results in microbiome data. *Biometrics*, pages 1–13, 2021.

[242] A. Thomas, G.K. Vishwakarma, and A. Bhattacharjee. Joint modeling of longitudinal and time-to-event data on multivariate protein biomarkers. *J. Comput. Appl. Math.*, 381:113016, 2020.

[243] A. Bhattacharjee, G.K. Vishwakarma, and A. Thomas. Bayesian state-space modeling in gene expression data analysis: An application with biomarker prediction. *Math. Biosci.*, 305:96–101, 2018.

[244] O. Rehman, H. Zhuang, A. Muhamed Ali, A. Ibrahim, and Z. Li. Validation of mirnas as breast cancer biomarkers with a machine learning approach. *Cancers*, 11(3):431, 2019.

[245] X. Fan, L. Shi, H. Fang, Y. Cheng, R. Perkins, and W. Tong. DNA microarrays are predictive of cancer prognosis: a re-evaluation. *Clin. Cancer Res.*, 16(2):629–636, 2010.

[246] T. Emura, S. Matsui, and V. Rondeau. *Survival analysis with correlated endpoints: joint frailty-copula models.* Springer, 2019.

[247] Karel GM Moons, Andre Pascal Kengne, Mark Woodward, Patrick Royston, Yvonne Vergouwe, Douglas G Altman, and Diederick E Grobbee. Risk prediction models: I. development, internal validation, and assessing the incremental value of a new (bio) marker. *Heart*, 98(9):683–690, 2012.

[248] Il Do Ha, John-Hyeon Jeong, and Youngjo Lee. *Statistical modelling of survival data with random effects*. Springer, 2017.

[249] John P Klein, Hans C Van Houwelingen, Joseph G Ibrahim, and Thomas H Scheike. *Handbook of survival analysis*. CRC Press, 2016.

[250] Eoghan R Malone, Marc Oliva, Peter JB Sabatini, Tracy L Stockley, and Lillian L Siu. Molecular profiling for precision cancer therapies. *Genome medicine*, 12(1):1–19, 2020.

[251] Lynda Chin and Joe W Gray. Translating insights from the cancer genome into clinical practice. *Nature*, 452(7187):553–563, 2008.

[252] Yinglei Lai, Baolin Wu, Liang Chen, and Hongyu Zhao. A statistical method for identifying differential gene–gene co-expression patterns. *Bioinformatics*, 20(17):3146–3155, 2004.

[253] Gareth James, Daniela Witten, Trevor Hastie, and Robert Tibshirani. *An introduction to statistical learning*, volume 112. Springer, 2013.

[254] Bojan Mihaljević, Ruth Benavides-Piccione, Concha Bielza, Javier DeFelipe, and Pedro Larrañaga. Bayesian network classifiers for categorizing cortical gabaergic interneurons. *Neuroinformatics*, 13(2):193–208, 2015.

[255] Konstantina Kourou, Themis P Exarchos, Konstantinos P Exarchos, Michalis V Karamouzis, and Dimitrios I Fotiadis. Machine learning applications in cancer prognosis and prediction. *Computational and Structural Biotechnology Journal*, 13:8–17, 2015.

[256] Alan Le Goallec and Chirag J Patel. Age-dependent co-dependency structure of biomarkers in the general population of the united states. *Aging (Albany NY)*, 11(5):1404, 2019.

[257] Ahmed M Alaa and Mihaela van der Schaar. Prognostication and risk factors for cystic fibrosis via automated machine learning. *Scientific Reports*, 8(1):1–19, 2018.

[258] Parham Aram, Lea Trela-Larsen, Adrian Sayers, Andrew F Hills, Ashley W Blom, Eugene V McCloskey, Visakan Kadirkamanathan, and Jeremy M Wilkinson. Estimating an individual's probability of revision surgery after knee replacement: a comparison of modeling approaches using a national data set. *American Journal of Epidemiology*, 187(10):2252–2262, 2018.

[259] Hong Wang and Gang Li. Extreme learning machine cox model for high-dimensional survival analysis. *Statistics in Medicine*, 38(12):2139–2156, 2019.

[260] Gajendra K Vishwakarma, Atanu Bhattacharjee, Souvik Banerjee, and Benoit Liquet. Classification algorithm for high-dimensional protein markers in time-course data. *Statistics in Medicine*, 39(28):4201–4217, 2020.

[261] Atanu Bhattacharjee, Jacinth Rajendra, Rajesh Dikshit, and Shilpee Dutt. Her2 borderline is a negative prognostic factor for primary malignant breast cancer. *Breast Cancer Research and Treatment*, 181(1):225–231, 2020.

[262] Benjamin Hofner, Andreas Mayr, Nikolay Robinzonov, and Matthias Schmid. Model-based boosting in r: a hands-on tutorial using the r package mboost. *Computational Statistics*, 29(1):3–35, 2014.

[263] Raphael Sonabend, Franz J Király, Andreas Bender, Bernd Bischl, and Michel Lang. mlr3proba: An r package for machine learning in survival analysis. *Bioinformatics*, 37(17):2789–2791, 2021.

[264] M Keller, J Kollmann, and PJ Edwards. Genetic introgression from distant provenances reduces fitness in local weed populations. *Journal of applied ecology*, 37(4):647–659, 2000.

[265] Nathan Keyfitz and Gary Littman. Mortality in a heterogeneous population. *Population studies*, 33(2):333–342, 1979.

[266] David Oakes. Bivariate survival models induced by frailties. *Journal of the American Statistical Association*, 84(406):487–493, 1989.

Index